Molecular Modeling Workbook

Warren J. Hehre
Wavefunction, Inc.

Alan J. Shusterman
Reed College

Janet E. Nelson
Wavefunction, Inc.

Organic Chemistry

Fourth Edition

L.G. WADE, JR.

Wavefunction, Inc.
18401 Von Karman Avenue, Suite 370
Irvine, CA 92612

PRENTICE HALL
Upper Saddle River, NJ 07458

Senior Editor: John Challice
Associate Editor: Kristen Kaiser
Editorial Assistant: Gillian Buonano
Special Projects Manager: Barbara A. Murray
Manufacturing Manager: Trudy Pisciotti
Supplement Cover Manager: Paul Gourhan
Supplement Cover Designer: PM Workshop Inc.

Upper Saddle River, NJ 07458

Printed in the United States of America

10 9 8 7 6 5 4 3 2 1

ISBN 0-13-030432-8

Prentice-Hall International (UK) Limited, London
Prentice-Hall of Australia Pty. Limited, Sydney
Prentice-Hall Canada, Inc., Toronto
Prentice-Hall Hispanoamericana, S.A., Mexico
Prentice-Hall of India Private Limited, New Delhi
Pearson Education Asia Pte. Ltd., Singapore
Prentice-Hall of Japan, Inc., Tokyo
Editora Prentice-Hall do Brazil, Ltda., Rio de Janeiro

To the Teacher

This book, a companion to Wade's **Organic Chemistry**, allows you to introduce modern molecular modeling into your organic chemistry course. The heart of the book is a CD-ROM containing over two hundred models. These span all of the topics routinely encountered in a two-semester science majors course. The CD-ROM also contains the SPARTANView program for Mac and PC compatible personal computers. Students can master its use after just a few minutes of training. Therefore, the complete package lets students (and teachers) focus on the most important job: learning organic chemistry.

The book consists of a number of sections: essays describing how to work with molecular modeling data, a tutorial describing the use of SPARTANView, over 50 organic chemistry problems which can only be solved using molecular models, and a tutorial describing the use of SPARTANBuild, an electronic model kit.

SPARTANView is needed to access the models on the CD-ROM, to query molecular structures, energies and other properties and to display a number of informative graphical surfaces. The tutorial provides a quick survey of its capabilities and intended manner of use.

The essays discuss the use of calculated energies to obtain thermochemical and kinetic data, and interpret the different graphical displays associated with the models on the CD-ROM (molecular orbitals, electron densities and electrostatic potentials).

The problems make up the bulk of this book. Each problem is presented on a single page. This comprises background chemistry, experimental observations, and a series of questions to be answered. The questions are intended to complement those found in Wade's **Organic Chemistry**. Thus, depending on the subject material, a student might be asked to draw Lewis structures that describe a molecule's structure, predict or interpret some aspect of molecular structure, identify a reactive site within a molecule, or to compare and rank molecules by their reactivity. The one thing that all of the questions have in common is that they can only be answered by examining the models that are provided. Therefore, problem-solving is tightly integrated with structure visualization. Students must look at and manipulate molecules in ways that they will not do otherwise.

Of course, molecular models are not derived from experiments, but rather from computer calculations. Thus, there will be some differences between modeling data and experimental data, and one must occasionally interpret these data in different ways. Nearly all of the models used in the workbook were calculated with SPARTAN using standard Hartree-Fock methods and the 3-21G basis set. This level of theory is of "intermediate" reliability.

The reliability and meaning of model energies requires special consideration. Model energies give the best results when used to compare molecules that contain the same number and type of chemical bonds, e.g., conformational isomers or the reactants and products in certain types of reactions. Even then, energy differences obtained from models may differ substantially from the free energy differences (ΔG) measured in solution that organic chemists are accustomed to. The total energies provided in this book are closely related to experimental enthalpies (ΔH). These provide an excellent starting point for describing gas-phase chemistry, but in order to convert them into accurate solution-phase ΔG values, corrections for interactions with solvent and for entropy would need to be made. This lies outside the scope of an elementary organic chemistry course. Therefore, we have limited energy calculations to situations where gas-phase energies provide useful qualitative (and often quantitative) estimates of solution-phase chemistry, and the data should be interpreted in this light.

SPARTANView and the model archive can be used separately from the book, thereby promoting a much wider range of model-based instruction. Many of the models in the archive have been chosen because of their utility as visual aids in chemistry lectures. A teacher with access to computer projection hardware can use SPARTANView and the models supplied to enhance virtually any lecture with easily manipulated, three-dimensional molecular models and animations. Using SPARTAN, teachers can also create their own models, and present them in class using SPARTANView. Teacher-generated models can also be transferred electronically to students so that they can be studied at the students' convenience. Thus, this book and its CD-ROM are the "starter kit" for introducing molecular modeling into organic chemistry.

The final section of the book concerns the use of SpartanBuild, an "electronic model kit" intended to replace the plastic models widely used by organic chemistry students. It offers a number of significant advantages. First, it allows building not only "normal" molecules, but also highly-strained systems, e.g, cyclopropane, as well as a variety of "unusual" molecules, e.g., reactive intermediates. Second, it accesses bond lengths and angles, allowing quantitative comparisons of molecular structures. Third, it provides for a variety of model display styles from skeletal representations which depict bonding to space-filling models which depict overall size and shape. Finally, it provides for structure refinement and reporting of strain energies.

A simple tutorial is provided, but no problems which make use of SpartanBuild have been supplied.

To the Student

Organic chemistry is a blend of the general and the specific. For example, organic chemists describe the oxygen-hydrogen (OH) bond as a "polar covalent bond." This description is quite general and valid for virtually every "OH"-containing molecule in existence, and it turns out that many "OH" molecules share common characteristics that can be attributed to this peculiar bond. The generalities tend to fall apart, though, when applied to specific molecules. Methanol, CH_3–OH, and octanol, $CH_3CH_2CH_2CH_2CH_2CH_2CH_2CH_2$–OH, both contain a polar covalent OH bond, but while this bond can make any number of methanol molecules dissolve in water, octanol is insoluble.

Your study of organic chemistry must integrate the general with the specific. You must not only learn general patterns but also how to apply them to specific molecules, and you must also learn the behavior of specific molecules in order to see where patterns come from. These skills can be learned in a variety of ways, but one of the most effective learning techniques is to study models of molecules that duplicate their size, shape, stability, and other chemically important properties. That is where this molecular modeling comes in.

This supplement to the fourth edition of Wade's **Organic Chemistry** contains over 50 problems that will allow you to build and refine your understanding of chemistry from the molecule's "eye view." This is achieved by basing every problem on a set of molecular models that you view and manipulate on your own personal computer. We believe that this combination of problems+models will improve your understanding of molecular structure and the relationship between molecular structure and molecular properties and chemical reactivity. More importantly, we believe that when you do the problems in this book you will gain a much better grasp of the **conceptual** basis of organic chemistry, and that this will make your study of organic chemistry more satisfactory and ultimately more successful.

This book is simple to use. Begin by loading the software and models onto your computer (see instructions on CD-ROM). Then read the tutorial describing the use of the SPARTANView program (this is the program used to access all of the models), and perform the instructions on your computer **as you read**. This last point needs to be emphasized. The tutorial and the problems can only be completed by working at your computer. Depending on your background, you may need to read some or all of the essays that describe how to work with modeling data (this information quickly becomes second-nature, especially if you make working these problems part of your regular study routine). Then tackle the problems. Be prepared to **look** at molecules, to move them, to measure them, to animate them, to find regions that are electron poor or electron rich, and most of all, to think about their chemistry. We guarantee that, after this, you will never look at molecules the same way again!

Table of Contents

I. Molecular Modeling in Organic Chemistry

Why is it important to introduce molecular modeling in the beginning organic chemistry course? With so many new concepts already essential to understanding organic chemistry, and with the mass of unfamiliar material already heaped upon the student, how can introduction of yet another dimension to the subject be justified? And, isn't modeling supposed to be grounded in quantum mechanics, the rudiments of which haven't even yet been presented to the student? Wouldn't it really be better to postpone consideration of molecular modeling until the basics are in place? We think not. Molecular modeling allows the student to think more clearly about issues which are fundamental to the study of organic chemistry – structure, stability and reactivity – than would be possible without the use of a computer.

In order to fully appreciate the widespread application that molecular modeling can find in beginning organic chemistry, it is important to appreciate the fundamental relationship between molecular structure and chemical, physical and biological properties and chemical reactivity. So-called structure-property and structure-reactivity relationships are explored in nearly every college chemistry course, whether introductory or advanced. Students are first taught about the structures of molecules, and are then taught how to relate structure to molecular properties and chemical reactivity.

The widespread use of this teaching technique, and the critical and central role of structural concepts in chemistry, suggests that the depiction and manipulation of structural models is a highly developed science. Unfortunately, this is not the case. The two-dimensional line figures, introduced more than a century ago to draw molecular structures, are still routinely used in education and research. Although easy for an expert to understand and produce, such drawings do not look at all like the molecules they are supposed to depict. In fact, learning how to interpret and create simple line drawings is one of the largest hurdles that students face, and is one of the principal reasons why many students find organic chemistry difficult.

Using computers to display molecular structure is an attractive alternative to traditional line drawings for several reasons. First, the model displayed on a computer screen "looks" and "behaves" more like a "real molecule" than a drawing does. The computer model can be viewed from different angles, and different display formats can be used to show atomic positions, atomic volumes, and other features of interest. Second, the computer can produce a good model even when the student does not know how to make an accurate drawing. Thus, the student, working with a computer, can explore "new areas of chemistry" where his or her knowledge of structure may be limited. Third, many molecules commonly encountered in beginning organic chemistry cannot be represented accurately, if at all, by simple drawings. These include molecules in which the charge is delocalized, many unstable molecules and, perhaps most important, reaction transition states. Computer models treat such species no differently than they handle structures which are well represented by conventional drawings. Fourth, molecular modeling can also be used to predict and display a variety of chemical and physical properties such as energy, dipole moment, and so on. Thus, the computer can be more than a simple structure display tool; it can also provide a means for visualizing, investigating and studying a multitude of chemical phenomena. These many advantages imply that the classroom use of computer modeling can be of enormous benefit in teaching about molecular structure, molecular properties and chemical reactivity.

Introduction of molecular modeling very early into the curriculum need not complicate or confuse the learning of organic chemistry, but rather assist the student in visualizing the structures of organic molecules and in learning the intimate connections between molecular structure and molecular properties and chemical reactivity.

II. Background for Molecular Modeling

This section comprises a series of essays intended to provide background material on essential quantities associated with molecular models:

> How to Use Energies to Calculate Thermodynamic and Kinetic Data.
>
> Molecular Orbitals. Quantum Mechanics in Pictures.
>
> Electron Densities. The Sizes and Shapes of Molecules.
>
> Electrostatic Potential Maps. Molecular Charge Distributions.

The first essay defines energy, relates it to more familiar heats of formation, and then goes on to show how calculated energies may be used to obtain heats of reactions ("thermodynamics") and rates of reactions ("kinetics").

The last three essays describe graphical quantities associated with the models on CD-ROM: molecular orbitals, electron densities and electrostatic potential maps. Molecular orbitals are related to bonding, electron densities to overall size and shape, and electrostatic potential maps to charge distribution.

How to Use Energies to Calculate Thermodynamic and Kinetic Data

In addition to molecular geometry, the most important quantity to come out of molecular modeling is the energy. Energy can be used to reveal which of several isomers is most stable, to determine whether a particular chemical reaction will have a thermodynamic driving force (an "exothermic" reaction) or be thermodynamically uphill (an "endothermic" reaction), and to ascertain how fast a reaction is likely to proceed. Other molecular properties, such as the dipole moment, are also important, but energy plays a special role.

There are many ways to express the energy of a molecule. Most common to organic chemists is as a **heat of formation, ΔH_f**. This is the heat of a hypothetical chemical reaction (it cannot be directly measured) that creates a molecule from so-called "standard states" of each of its constituent elements. For example, the heat of formation for methane would be the energy required to create CH_4 from graphite and H_2, the "standard states" of carbon and hydrogen, respectively.

An alternative, **total energy**, will be used throughout this workbook. The total energy is the heat of a hypothetical reaction that creates a molecule from a collection of separated nuclei and electrons. Like the heat of formation, total energy cannot be measured directly, and is used solely to provide a standard method for expressing and comparing energies.

Total energies are always negative numbers and, in comparison with the energies of chemical bonds, are very large. They are generally expressed in "so-called" **atomic units** or **au**, but may be converted to other units as desired:

$$1 \text{ au} = 627.5 \text{ kcal/mol} = 2625 \text{ kJ/mol}$$

Total energies (like heats of formation) may be used to calculate energies of **balanced** chemical reactions (reactants → products):

$$\Delta E(\text{reaction}) = E_{\text{product1}} + E_{\text{product2}} + \ldots - E_{\text{reactant1}} - E_{\text{reactant2}} - \ldots$$

A negative ΔE indicates an exothermic (thermodynamically favorable) reaction, while a positive ΔE an endothermic (thermodynamically unfavorable) reaction.

Comparison of isomer stability is a special case. It involves a chemical reaction in which the "reactant" is one isomer and the "product" is another isomer (isomer1 $\rightarrow$ isomer2).

$$\Delta E(\text{isomer}) = E_{\text{isomer2}} - E_{\text{isomer1}}$$

A negative ΔE means that isomer2 is more stable than isomer1, and a positive ΔE means that isomer2 is less stable than isomer 1.

Total energies may also be used to calculate activation energies, $\Delta E^{\ddagger}$.

$$\Delta E^{\ddagger} = E_{\text{transition state}} - E_{\text{reactant1}} - E_{\text{reactant2}} - \ldots$$

Here, $E_{\text{transition state}}$ is the total energy of the transition state, and $E_{\text{reactant1}}$, ... are the total energies of the reactants.

Although there are many situations where one needs only to know whether a reaction is exothermic or endothermic, or if one reaction is more exothermic than another, there are other situations where energies need to be converted into other quantities, in particular equilibrium constants, reaction rate constants and half lives.

Equilibrium concentrations of reactants and products can be calculated from the equilibrium constant, K_{eq}, which is related to the free energy of reaction, ΔG_{rxn}:

$$K_{eq} = \exp(-\Delta G_{rxn}/RT)$$

Here R is the gas constant and T is the temperature (in K). At room temperature (298K) and for ΔG_{rxn} in au, this is given by:

$$K_{eq} = \exp(-1060\ \Delta G_{rxn})$$

ΔG_{rxn} has two components, the enthalpy of reaction, ΔH_{rxn}, and the entropy of reaction, ΔS_{rxn}. These are defined by the following formulas:

$$\Delta G_{rxn} = \Delta H_{rxn} - T\Delta S_{rxn}$$

$$\Delta H_{rxn} \approx \Delta E_{rxn} = E_{product1} + E_{product2} + \ldots - E_{reactant1} - E_{reactant2} - \ldots$$

$$\Delta S_{rxn} = S_{product1} + S_{product2} + \ldots - S_{reactant1} - S_{reactant2} - \ldots$$

Although ΔG_{rxn} depends on both enthalpy and entropy, there are many reactions for which the entropy contribution is small, and can be neglected. Thus, if $\Delta H_{rxn} \approx \Delta E_{rxn}$, equilibrium constants for such reactions may be estimated by the following equation:

$$K_{eq} \approx \exp(-\Delta E_{rxn}/RT) \approx \exp(-1060\ \Delta E_{rxn})$$

Reaction rate constants, k_{rxn}, are also related to free energies. As before, if entropy contributions can be neglected, the rate constant can be obtained directly from the activation energy, $\Delta E^{\ddagger}$, by:

$$k_{rxn} \approx (k_B T/h)[\exp(-\Delta E^{\ddagger}/RT)]$$

Here k_B and h are the Boltzmann and Planck constants, respectively. At room temperature and for $\Delta E^{\ddagger}$ in au, k_{rxn} is given by:

$$k_{rxn} = 6.2\mathrm{x}10^{12} \exp(-1060\ \Delta E^{\ddagger})$$

Another way to describe reaction rates is by half-life, $t_{1/2}$, the amount of time it takes for the reactant concentration to drop to one half of its original value. When the reaction follows a first-order rate law, rate = $-k_{rxn}$[reactant], $t_{1/2}$ is given by:

$$t_{1/2} = \ln 2/k_{rxn} = 0.69/k_{rxn}$$

Molecular Orbitals. Quantum Mechanics in Pictures

Among the methods that chemists have developed for describing electrons in molecules, Lewis structures are perhaps the most familiar. These drawings assign electrons either to single atoms (lone pairs) or pairs of atoms (bonds).

An alternative method for describing electron motion in molecules is the method of molecular orbitals. You are probably already familiar with atomic orbitals. These are actually the mathematical solutions to the quantum mechanical equations that describe electron motion inside atoms. The orbitals resemble waves in that they typically have large positive magnitudes in some regions of space (a "crest"), have large negative magnitudes in others (a "trough"), and pass through zero, or vanish, somewhere in between ("go through a node"). Molecular orbitals are defined and calculated in the same way as atomic orbitals and they display similar wave-like properties. The main difference between molecular and atomic orbitals is that molecular orbitals are not confined to a single atom. The "crests" and "troughs" in an atomic orbital are confined to a region close to the atomic nucleus (typically within 1-2 Å). The electrons in a molecule, on the other hand, do not "stick" to a single atom, and are free to move all around the molecule. Consequently, the "crests" and "troughs" in a molecular orbital are usually spread over several atoms.

Orbital Surfaces. Molecular orbitals provide important clues about chemical reactivity, but before this information can be used, it is necessary to understand what molecular orbitals look like. The following figure shows two representations, a drawing and a computer-generated picture, of an unoccupied molecular orbital of hydrogen molecule, H_2.

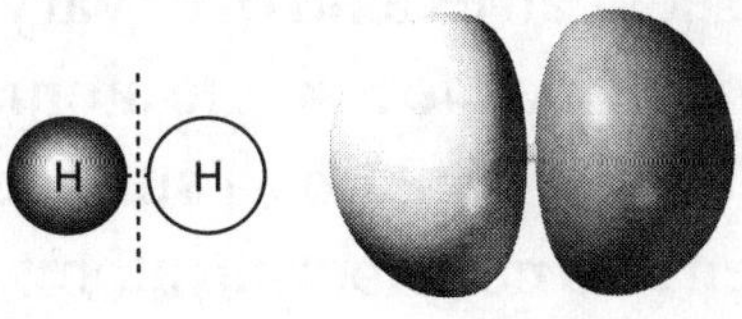

The drawing superimposes two circles and a dashed line on the molecule. The circles identify regions of space where the orbital takes on a significant value, either positive (shaded circle) or negative (unshaded circle). The dashed line identifies an orbital node, locations where the orbital's value is exactly zero. The drawing is useful, but it is also limited. We only obtain information about the orbital in two dimensions, and we only learn the location of "significant" regions and not how the orbital builds and decays inside and outside of these regions.

The computer-generated picture shows an "orbital surface." The surface is mathematically accurate in that it is derived from an authentic (but approximate) calculated solution to the quantum mechanical equations of electron motion. The surface is three-dimensional. It can be manipulated using a computer, and can be looked at from a variety of different perspectives. Note that the "orbital surface" actually consists of two distinct surfaces represented by different colors. The surfaces have the same meaning as the two circles in the drawing. They identify regions where the orbital takes on a significant value, either positive (blue) or negative (red). The orbital node is not shown, but we can guess that it must lie midway between the two surfaces (this follows from the fact that the orbital's value can only change from positive to negative by passing through zero).

Orbital Shapes and Chemical Bonds. Although molecular orbitals and Lewis structures are both used to describe electron distributions in molecules, they are used for different purposes. Lewis structures are used to count the number of bonding and nonbonding electrons around each atom. Molecular orbitals are not useful as counting tools, but orbital shapes are useful tools for describing chemical bonding and reactivity. This section describes a number of common orbital shapes and illustrates how they may be used to interpret chemical bonding and reactivity.

Molecular orbital surfaces can extend over varying numbers of atoms. If the orbital surface is confined to a single atom, the orbital is regarded as nonbonding. If the orbital contains a surface that extends over the bonding region between two neighboring atoms, the orbital is regarded as bonding with respect to these atoms. Adding electrons to this orbital will strengthen the bond between these atoms and cause them to draw closer together, while removing electrons will have the opposite effect. The following pictures show drawings and orbital surfaces for two different kinds of bonding orbitals. The drawing and surface on the left correspond to a σ bond while the drawing and surface on the right correspond to a π bond.

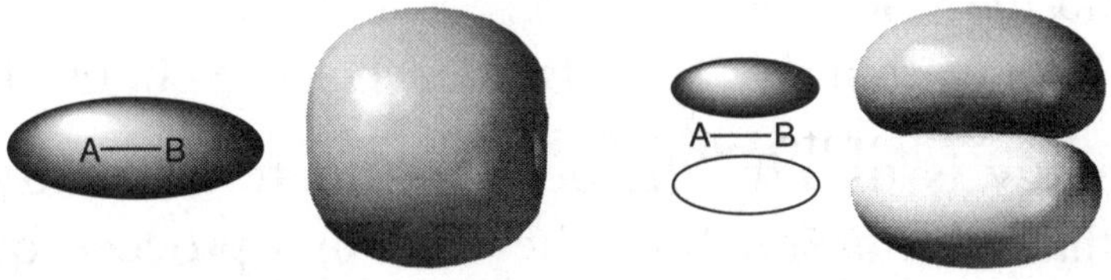

σ bond π bond

It is also possible for an orbital to contain a node that divides the bonding region into separate "atomic" regions. This orbital is regarded as antibonding with respect to these atoms. Adding electrons to an antibonding orbital weakens the bond and drives the atoms apart, while removing electrons from the orbital has the opposite effect. The following pictures show drawings and orbital surfaces for σ (left) and π (right) antibonds.

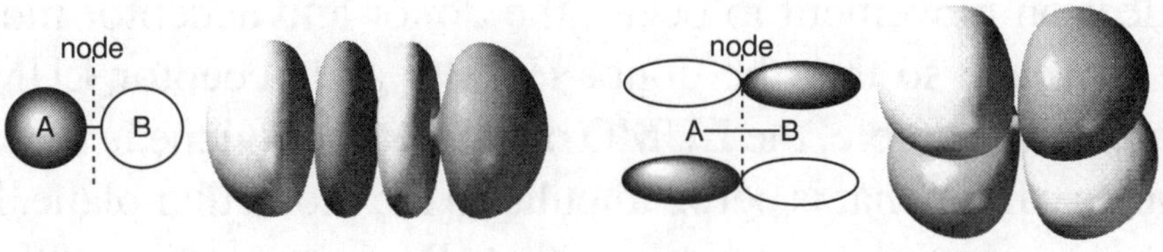

σ antibond π antibond

Notice that bonds can be strengthened in two different ways, by adding electrons to bonding orbitals, and by removing electrons from antibonding orbitals. The converse also holds.

Frontier Orbitals and Chemical Reactivity. Chemical reactions typically involve movement of electrons from an "electron donor" (base, nucleophile, reducing agent) to an "electron acceptor" (acid, electrophile, oxidizing agent). This electron movement between molecules can also be thought of as electron movement between molecular orbitals. It follows that the properties of these "electron donor" and "electron acceptor" orbitals provide considerable insight into chemical reactivity.

The first step in constructing a "molecular orbital" picture of a chemical reaction is to decide which orbitals are most likely to serve as the "electron donor" and "electron acceptor" orbitals. It should be obvious that the "electron donor" orbital must be drawn from the set of occupied orbitals, and the "electron acceptor" orbital must be an unoccupied orbital, but there are many orbitals in each set to choose from.

Orbital energy is usually the deciding factor. Usually chemical reactions that are observed are the ones that proceed quickly, and such reactions typically have small energy barriers. Therefore, chemical reactivity should be associated with the donor-acceptor orbital combination that requires the smallest energy input for electron movement. The best combination is typically the one involving the highest-energy occupied molecular orbital (or HOMO) as the donor orbital and the lowest-energy unoccupied molecular orbital (or LUMO) as the acceptor orbital. The HOMO and LUMO are collectively referred to as the "frontier orbitals," and most chemical reactions involve electron movement between them.

For electron movement to occur, the donor and acceptor molecules must approach so that the donor HOMO and acceptor LUMO can interact. For example, the LUMO of singlet methylene is a 2p atomic orbital on carbon that is perpendicular to the molecular plane. Donors must approach methylene in a way that allows interaction of the donor HOMO with the 2p orbital.

LUMO of methylene (top) approaching HOMO of donor molecule (bottom)

In certain cases, multiple frontier orbital interactions must be considered, for example, in the Diels-Alder cycloaddition between 1,3-butadiene and ethene.

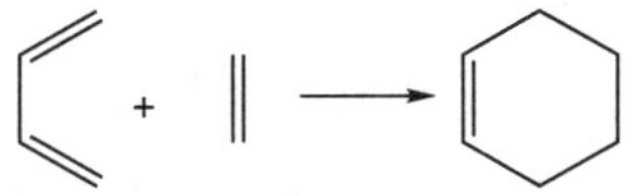

Here the reactants must combine in a way that allows two bonds to form simultaneously. This implies two different sites of satisfactory frontier orbital interaction. Focusing exclusively on interactions involving only the terminal carbons in each molecule, several different frontier orbital combinations are possible.

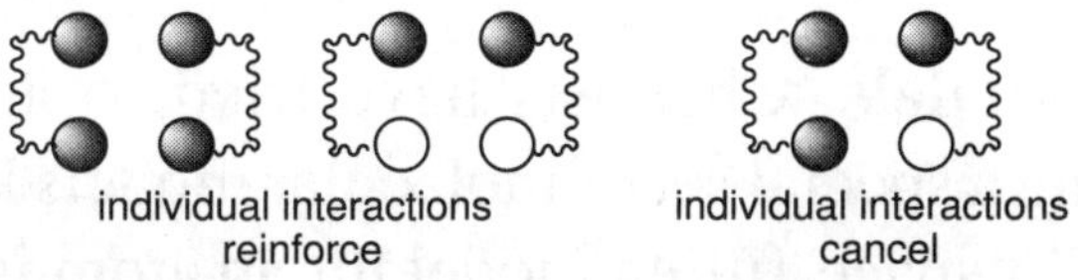

In all three combinations, the "upper" orbital components are the same sign, and their overlap is positive. In the two cases on the left, the lower orbital components also lead to positive overlap. Thus, the upper and lower interactions reinforce, and the total frontier orbital interaction is non-zero. Reaction can occur. The right-most case is different. Here the lower orbital components lead to negative overlap (the orbitals have opposite signs at the interacting sites), and the total overlap is zero. No reaction can occur in this case.

Frontier orbital interactions in the Diels-Alder cycloaddition are like those found in the middle drawing, i.e., the upper and lower interactions reinforce and the reaction proceeds. The cycloaddition of two ethene molecules (shown below), however, involves a frontier orbital interaction like the one on the right, so this reaction does not occur.

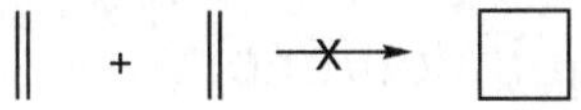

Electron Densities. The Sizes and Shapes of Molecules

How "big" is an atom or a molecule? It should be obvious that atoms and molecules take up space. A gas can be compressed into a smaller volume but only so far. Liquids and solids cannot be easily compressed. While the individual molecules in a gas are widely separated and can be pushed into a much smaller volume, the molecules in a liquid or a solid are already close together and cannot be "squeezed" much further. The "bottom line" is that atoms and molecules require a certain amount of space. But how much?

Space-Filling Models. Chemists have tried to answer the "size" question by using a special set of molecular models known as "space-filling" models. A space-filling model for an atom is simply a sphere of fixed radius. A different radius is used for each element, and the radii are chosen to reproduce certain experimental observables, such as the compressibility of a gas, or the spacing between atoms in a crystal. Space-filling models for molecules consist of a set of interpenetrating atomic spheres. This reflects the fact that chemical bonds that hold the molecule together cause the atoms to move very close together. The degree of "interpenetration" can be used as a criterion for chemical bonding. If two atomic spheres strongly interpenetrate then the atoms must be bonded. Otherwise the atoms are not bonded.

Space-filling models for ammonia, trimethylamine and quinuclidine show how "big" these molecules are. Ammonia is the smallest, and quinuclidine the biggest. The models also show that the nitrogen in ammonia is more "exposed" than the corresponding nitrogen atoms in trimethylamine and quinuclidine.

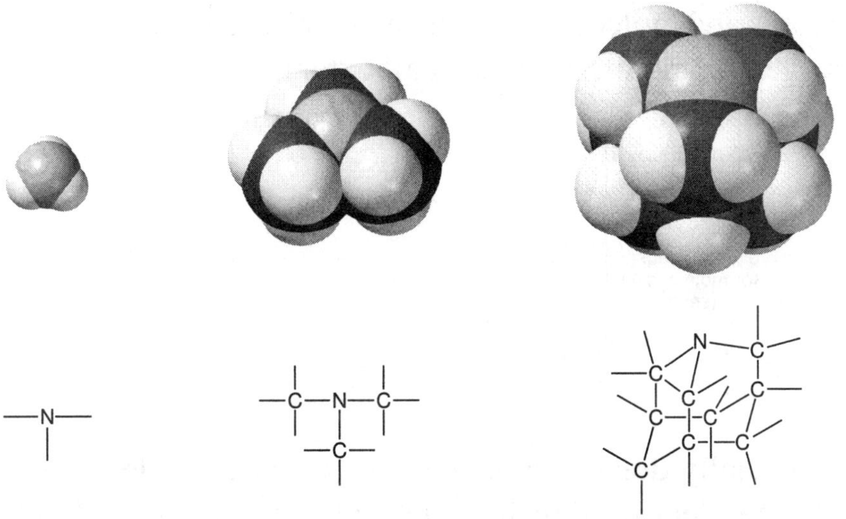

space-filling models of ammonia (left), trimethylamine (center) and quinuclidine (right)

Electron Density Surfaces. An alternative measure of molecular size and shape is the electron cloud. Atoms and molecules are made up of positively-charged nuclei surrounded by a negatively-charged electron cloud, and it is the size and shape of the electron cloud that defines the size and shape of an atom or molecule. Quantum mechanics provides the mathematical recipe for determining the size and shape of the electron cloud. The size and shape of an electron cloud is described by the "electron density" (the number of electrons per unit volume). Consider a graph of electron density in the hydrogen atom as a function of distance from the nucleus.

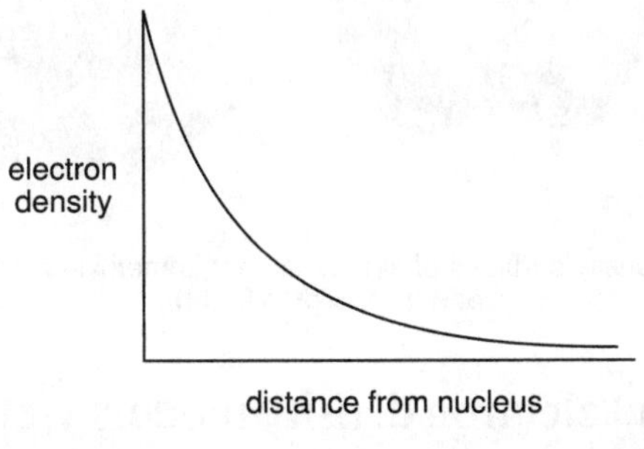

The graph brings up an important problem. Namely, the electron cloud lacks a clear boundary. While electron density decays rapidly with distance from the nucleus, nowhere does it fall to zero. Therefore, when atoms and molecules "rub up against each other," their electron clouds overlap and merge to a small extent.

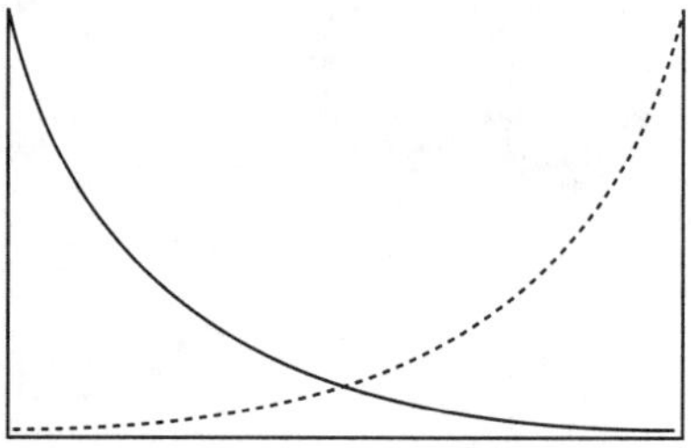

What is actually done in practice is to select a single ("boundary") value of the electron density, such that the electron cloud that it encloses is approximately the same size and shape as that of a conventional space-filling model. The resulting surface can be thought of as a molecule's "outer skin," and we can use the volume inside this surface to define molecular size. This approach is used throughout this book, but to simplify things we will abbreviate "outer skin electron density surface" to just "electron density surface."

The following picture shows that electron density surfaces for ammonia, trimethylamine and quinuclidine are qualitatively very similar to the space-filling models shown previously.

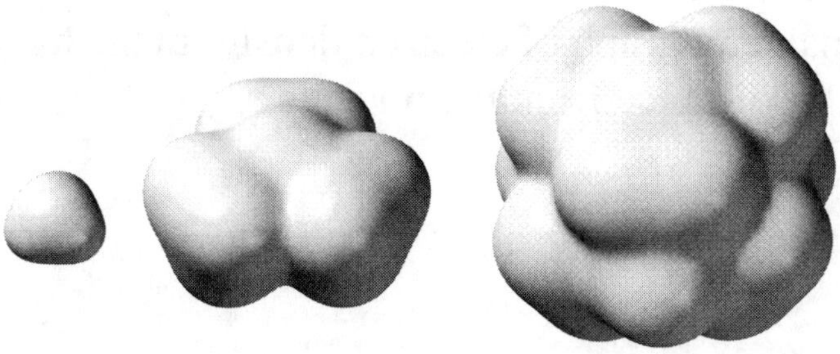

electron density surfaces of ammonia (left), trimethylamine (center) and quinuclidine (right)

Both space-filling and electron density models yield similar molecular volumes, and both show the obvious differences in overall size. Because the electron density surfaces provide no discernible boundaries between atoms (and employ no colors to highlight these boundaries), the surfaces may appear to be less informative than space-filling models in helping to decide to what extent a particular atom is "exposed." This "weakness" raises an important point, however. Electrons are associated with a molecule as a whole and not with individual atoms. The space-filling representation of a molecule in terms of discernible atoms does not reflect reality, but rather is an artifact of the model. The electron density surface is more "real" in that it shows a single electron cloud for the entire molecule.

Spin Density Surfaces. Electrons exist in either of two spin states: "spin up" or "spin down." Almost all of the molecules that you will encounter in your study of organic chemistry will involve each "spin-up" electron paired to a "spin down" electron. Thus, the number of "spin up" and "spin down" electrons will be the same, and the electron clouds due to each spin will be identical. There are some notable exceptions, most important being free radicals which contain an odd number of electrons. Since the number of "spin up" and "spin down" electrons cannot be equal, the "spin up" and "spin down" electron clouds cannot be identical.

The "spin density surface" is a tool which allows the unpaired electrons to be identified. "Spin density" is defined as the difference between the "spin up" and "spin down" electron clouds, and a spin density surface is constructed by connecting together points in the electron cloud where the spin density has an arbitrarily chosen value.

The usefulness of spin density surfaces can be seen in the following models of methyl radical, CH_3, and allyl radical, CH_2=$CHCH_2$. In each case, the surface is shaped somewhat like a 2p atomic orbital on carbon. While the unpaired electron is confined (localized) to the carbon atom in methyl radical, it is delocalized over the two terminal carbons in allyl radical.

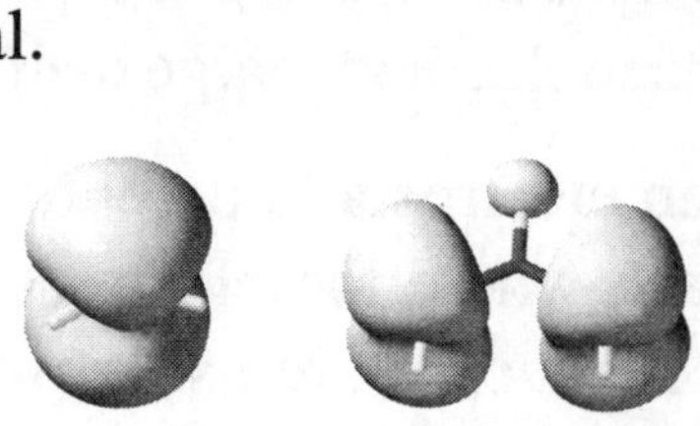

spin density surfaces of methyl radical (left) and allyl radical (right)

Electrostatic Potential Maps. Molecular Charge Distributions

The charge distribution in a molecule can provide critical insight into its physical and chemical properties. For example, organic molecules that are charged, or highly polar, tend to be water-soluble, and polar molecules may stick together in specific geometries, such as the "double helix" in DNA. Chemical reactions are also associated with charged sites, and the most highly charged molecule, or the most highly charged site in a molecule, is often the most reactive. The type of charge (positive or negative) is also important. Positively-charged sites in a molecule invite attack by bases and nucleophiles, while negatively-charged sites are usually targeted by acids and electrophiles.

One way to describe a molecule's charge distribution is to give a numerical "atomic charge" for each atom. Unfortunately, all of the available methods for assigning charge necessarily bias the calculated charges in one way or another. An alternative for describing molecular charge distributions makes use of a quantity termed the "electrostatic potential." This is defined as the energy of interaction of a point positive charge with the nuclei and electrons of a molecule, and its value depends on the location of the point positive charge. If the point charge is placed in a region of excess positive charge (an electron-poor region), the point charge-molecule interaction is repulsive and the electrostatic potential is positive. Conversely, if the point charge is placed in a region of excess negative charge (an electron-rich region), the interaction is attractive and the electrostatic potential is negative. Thus, by moving the point charge around the molecule, a "map" of the molecular charge distribution can be created.

It is possible to make an electrostatic potential "surface" by finding all of the points in space where the electrostatic potential matches some particular value. A much more useful way to show molecular charge distribution, however, is to construct a map that can show variation in electrostatic potential. This is normally done in two steps.

First one constructs the molecule's electron density surface or "outer skin" to define the locations being mapped. Then one constructs a map by using different colors to represent the different values of the electrostatic potential on this surface. Mapping requires an arbitrary choice for a color scale. Red, the low energy end of the spectrum, is used to color the regions of most negative (least positive) electrostatic potential, and blue is used to color the regions of most positive (least negative) electrostatic potential. Intermediate colors represent intermediate values of the electrostatic potential, so that potential increases in the order: red < orange < yellow < green < blue.

The connection between a molecule's electron density surface, an electrostatic potential surface, and the molecule's electrostatic potential map can be illustrated for benzene. An electron density surface defines molecular shape and size, much as a conventional space-filling model.

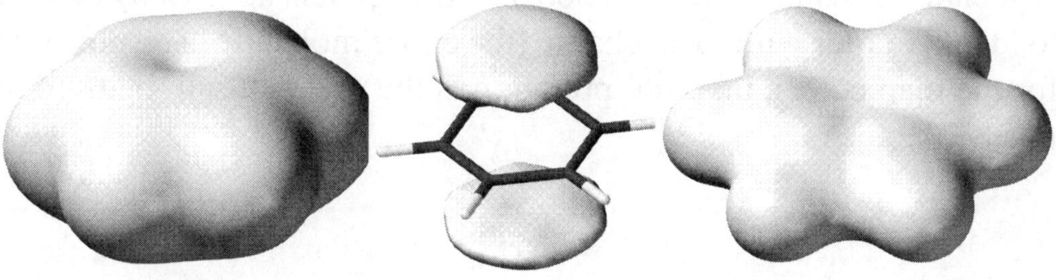

electron density surface for benzene (left), electrostatic potential surfaces for benzene: –.03 au (middle) and +0.3 au (right)

An electrostatic potential surface corresponding to points where the potential equals –.03 au (≈ –20 kcal/mol) comprises two parts, one above the face of the ring and the other below. Since the molecule's π electrons lie closest to these surfaces, it may be concluded that these electrons can attract a point positive charge (or an electrophile) to the molecule. A "positive" electrostatic potential surface corresponding to points where the potential equals +.03 au (~+20 kcal/mol) has a completely different shape. It is disk-shaped and wrapped fairly tightly around the nuclei. The shape and location of this surface indicates that a point positive charge is repelled to this region, or that a point negative charge (a nucleophile) would be attracted here.

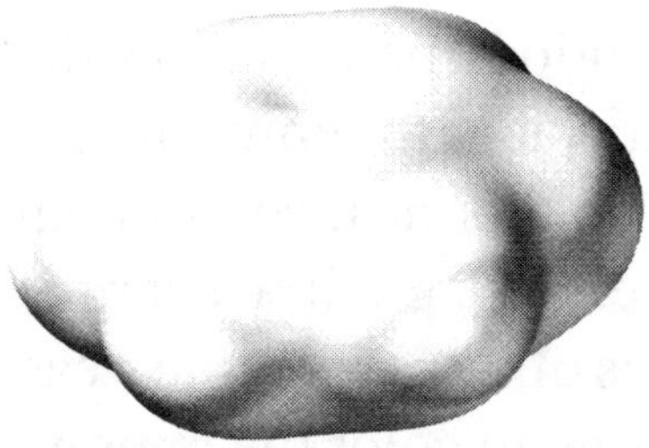

electrostatic potential map for benzene

The electrostatic potential map of benzene conveys the molecule's size as well as its charge distribution in a much more compact manner. The size and shape of the map are, of course, identical to that of the electron density surface, and indicate what part of the molecule is easily accessible to other molecules. The colors reveal the overall charge distribution. The faces of the ring, the π system, are "red" (electron rich), while the plane of the molecule (and the hydrogens especially) is "blue" (electron poor). Although it is not strictly correct to identify color with local charge (the entire molecule is responsible for the map color), this interpretation is the one that will be used in this book.

III. How to Use SPARTANView

This section serves as a practical introduction to the SPARTANView program for Power Mac's and PC's (Windows 95/98/NT). It will show you how to: 1) view and manipulate molecules on screen, 2) measure bond distances, angles and dihedral angles, 3) display energies, dipole moments, atomic charges and frequencies and 4) display molecular orbitals, electron and spin densities and electrostatic potential maps.

File Menu: Opening, Closing and Manipulating Molecules

The **File** menu accesses the molecule archive.

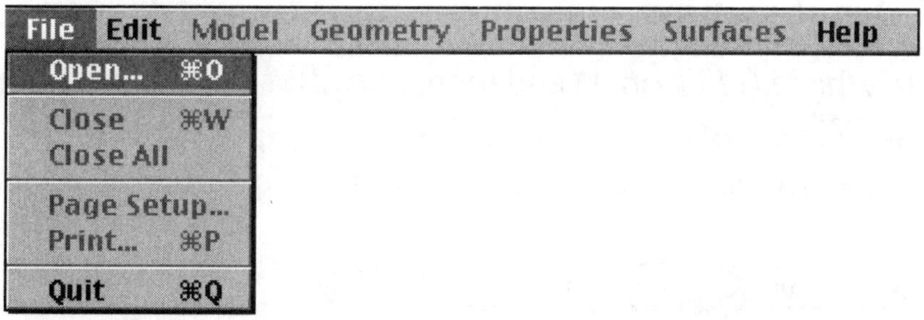

Open "**Tutorial**." This brings all the molecules discussed in this section onto the screen. All molecules can be removed by selecting **Close All.** Individual molecules may be removed by first selecting them (see below) and then selecting **Close**.

The mouse, together with one or more keys, is used both to select and manipulate (translate, rotate and scale) molecules. Available functions are listed below.

MAC	
Select	click on object
Rotate	move mouse with button depressed
Translate	press **option key** and move mouse with button depressed
Scale	press **option key** and ⌘ together and move mouse with button depressed

PC	
Select	click (left mouse button) on object
Rotate	move mouse with left button depressed
Translate	move mouse with right button depressed
Scale	press **shift key** and move mouse with right button depressed

Identify ***ethane*** on the screen, and *click* on it (left mouse button on the PC) to make it the selected molecule. Practice rotating and translating ethane. Select a different molecule, and then rotate and translate it.

Model Menu: Viewing Molecules with Different Models

Return to ethane (*click* on it) and then, one after the other, select **Wire**, **Ball and Wire**, **Tube**, **Ball and Spoke**, or **Space Filling** from the **Model** menu to view ethane with a variety of different models.

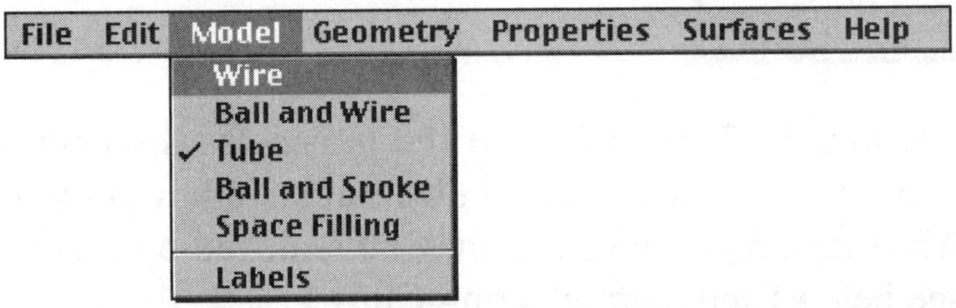

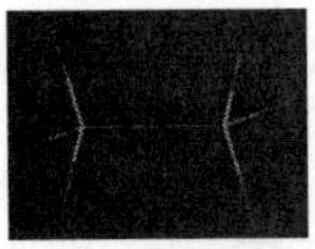
Wire

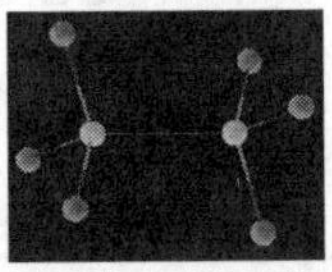
Ball and Wire

Tube

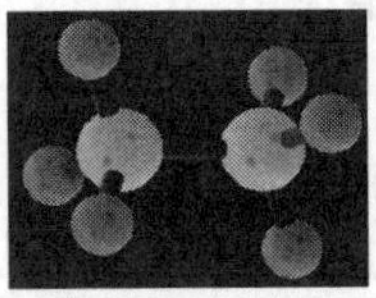
Ball and Spoke

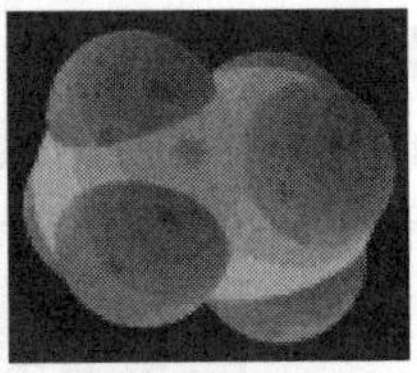
Space Filling

> The **3** key is used to toggle between stereo 3-D and regular display. To view in 3-D you will need to wear red/blue glasses.

The first four models for ethane show roughly the same information. The **Wire** model looks like a line formula in your chemistry textbook, except that all atoms, not just carbons, are found at the end of a line or at the intersection of lines. (The only exception occurs where three atoms lie on a line. Here, a **Wire** model will not show the exact position of the center atom.) The **Wire** model uses color to distinguish different atoms (see table below), and one, two and three lines to indicate single, double and triple bonds, respectively.

Atom Colors			
Hydrogen	white	Sodium	yellow
Lithium	tan	Silicon	grey
Beryllium	green	Phosphorous	tan
Boron	tan	Sulfur	sky blue
Carbon	black	Chlorine	tan
Nitrogen	blue-gray	Potassium	orange
Oxygen	red	Bromine	orange
Fluorine	green		

The **Ball and Wire** model is identical to the **Wire** model, except that atom positions are represented by small spheres. This makes it possible to identify all atom locations in all molecules. The **Tube** model is identical to the **Wire** model, except that bonds, whether single, double or triple, are represented by single colored tubes. The tubes are useful because they better convey the three-dimensional shape of a molecule. The **Ball and Spoke** model is a variation on the **Tube** model; atom positions are represented by colored spheres, making it possible to see all atom locations in all molecules.

The most novel model is the **Space-Filling** model. No bonds are shown. Rather, each atom is displayed as a colored sphere that represents the atom's approximate size, and the complete model indicates the molecule's approximate size. The existence (or absence) of bonds can be inferred from the amount of overlap between neighboring atomic spheres. If two spheres substantially overlap, then the atoms are almost certainly bonded, and conversely, if two spheres hardly overlap, then the atoms are not bonded. Intermediate overlaps suggest "weak bonding," for example, hydrogen bonding.

Atoms may be labelled by selecting **Labels** (labels may be "turned off" by selecting **Labels** a second time). Only wire and ball-and-wire models may be labelled.

Geometry Menu: Measuring Molecular Geometries

Distances, angles, and dihedral angles can easily be measured with SPARTANView using **Distance**, **Angle** and **Dihedral**, respectively, from the **Geometry** menu.

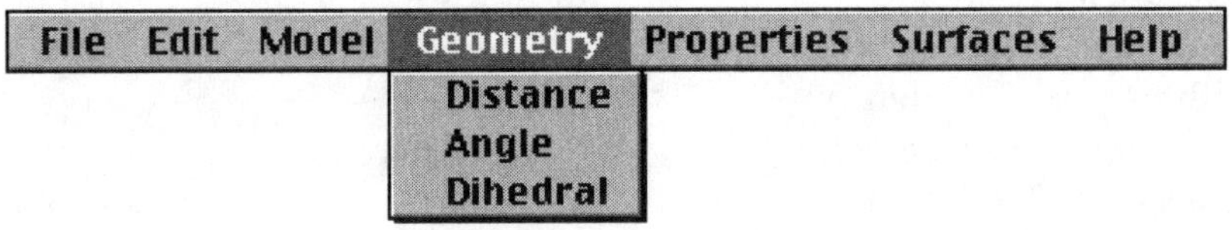

A. Distance: This measures the distance between two atoms. First select ***propene*** from the molecules on screen, and then select **Distance** from the **Geometry** menu. *Click* on a bond or on any two atoms (the atoms do not need to be bonded). The distance (in Ångstroms) will be displayed at the bottom of the screen. Repeat the process as necessary, and *click* on **Done** (at the bottom right of the screen) when finished.

B. Angle: This measures the angle around a central atom. Select ***ammonia*** from the molecules on screen, and then select **Angle**. *Click* first on H, then on N, then on another H, or on two NH bonds. The angle (in degrees) will be displayed at the bottom of the screen. Repeat the process as necessary, and *click* on **Done** (at the bottom right of the screen) when finished.

C. Dihedral: This measures the angle formed by two intersecting planes, the first plane containing the first three atoms and the second plane containing the last three atoms. Select ***hydrogen peroxide*** from the molecules on screen, and then select **Dihedral**. *Click* on the four atoms in the sequence H O O H (or the three bonds in sequence H–O O–O O–H). The dihedral angle (in degrees) will be displayed at the bottom of the screen. Repeat the process as necessary, and *click* on **Done** (at the bottom right of the screen) when finished.

Properties Menu: Displaying Molecular Properties

Energies, dipole moments, atomic charges and frequencies are available under the **Properties** menu.

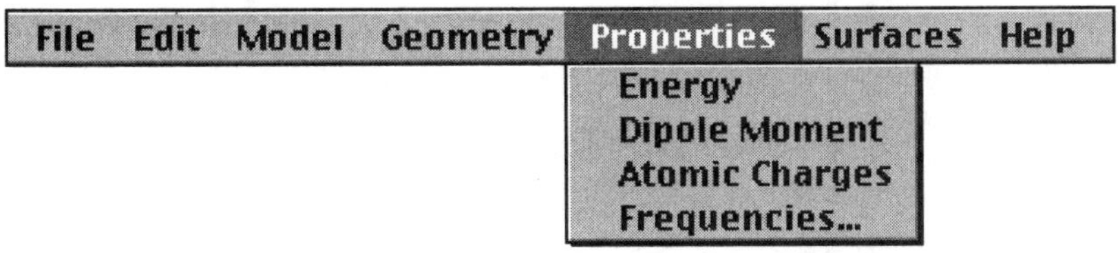

A. Energy: Select ***acetic acid*** from the molecules on screen, and then select **Energy** from the **Properties** menu. The energy of acetic acid (in atomic units or au) is displayed at the bottom of the screen. *Click* on **Done** (at the bottom right of the screen) when finished.

B. Dipole Moment: To display the dipole moment of acetic acid, select **Dipole Moment**. The magnitude of the dipole moment (in debyes) is displayed at the bottom of the screen and the dipole moment vector "+⟶", where "+" to "–" refer to the positive and negative ends of the dipole moment, respectively, is attached to the model on screen. *Click* on **Done** (at the bottom right of the screen) when finished.

C. Atomic Charges: To display atomic charges for acetic acid, select **Atomic Charges**. *Click* on an atom. The charge on that atom is displayed at the bottom of the screen. A positive number indicates a deficiency of electrons and a negative number, an excess of electrons. Repeat the process as necessary for different atoms, and *click* on **Done** (at the bottom right of the screen) when finished.

D. Frequencies: To display the molecular vibrations of ***water*** first display it as a ball-and-spoke model (**Ball and Spoke** from the **Model** menu), then select **Frequencies**, *double click* on a frequency (in cm^{-1}) in the dialog which results, and then *click* on **OK**. You can select another frequency by reentering the dialog, *double clicking* on another frequency and *clicking* on **OK**. You can turn off the animation by reentering the dialog, *double clicking* on the selected frequency and *clicking* on **OK**.

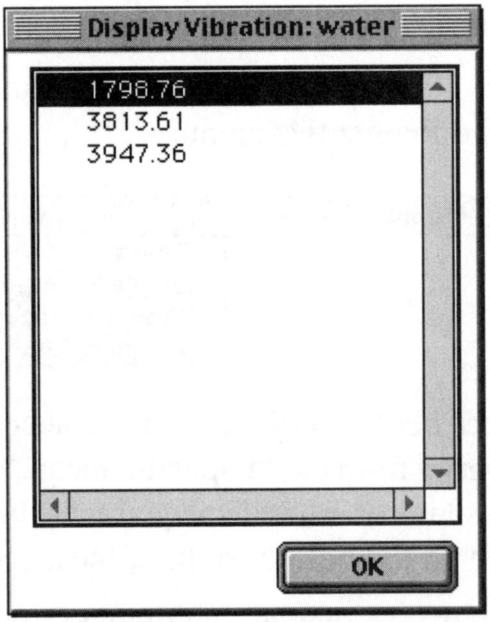

Surfaces Menu: Displaying Graphical Surfaces

Electron densities and spin densities, as well as particular molecular orbitals may be displayed as graphical surfaces. In addition, the value of the electrostatic potential may be mapped onto an electron density surface (an "electrostatic potential map") to show overall charge distribution.

> While SPARTANView allows you to display two (or more) surfaces or maps at one time, this can lead to confusion and is not advised. Be certain to "turn off" one surface before displaying another surface.

Surfaces and maps are accessible from the **Surfaces** menu.

This menu will contain one or more entries describing the available surfaces.

Select ***ethene*** from the molecules on screen. Select **Surfaces** and then **Solid** under the **HOMO** sub-menu.

This will result in the display of ethene's highest-occupied molecular orbital as a solid. It is a π orbital, equally concentrated above and below the plane of the molecule. The colors ("red" and "blue") give the sign of the orbital.

Select ***benzene*** from the molecules on screen, and select **Surfaces**. **Potential Map** refers to an electrostatic potential map. Select **Transparent** to allow you to see the molecular skeleton underneath. The surface is colored "red" in the π system (indicating that this region is attracted to a positive charge), and "blue" in the σ system (indicating that this region is repelled by a positive charge).

Collections of Molecules

Molecules can be grouped together and stepped through rapidly ("animated") to portray a conformational change or chemical reaction.

Select ***bromide+tert-butyl chloride*** from the molecules on screen. This provides a series of "frames" describing the S_N2 displacement of chloride in *tert*-butyl chloride by bromide. A bar appears at the bottom of the screen.

You can step forward or backward through the individual frames by *clicking* on ◀ and ▶, respectively, at the right of the bar or using the slider. You can animate the sequence of frames by *clicking* on ▶ at the left of the bar. The animation can be "turned off" by *clicking* on ⏸ (which replaces ▶) at the left of the bar. Practice these functions and pay particular attention to the changes in geometry which occur during the reaction. Experiment with different model types to get the clearest picture.

Select **Energy** (**Properties** menu). Notice that it updates automatically as you go from one frame to another. This allows you to easily construct reaction energy diagrams (energy vs. frame number or vs. a specific geometrical parameter). Make such a plot for this S_N2 reaction. Note, that the reaction as written is thermodynamically favorable, i.e., it is exothermic. Note also, that only a relatively small energy barrier needs to be surmounted.

Edit Menu: Copying Graphics and Data

Copy under the **Edit** menu is used for copying graphics and data onto the clipboard. This can later be imported into such programs as Microsoft Word and Excel. There are two modes of operation:

Where **Distance**, **Angle** or **Dihedral** (**Geometry** menu) or **Energy**, **Dipole Moment** or **Atomic Charges** (**Properties** menu) have been selected, **Copy** copies the selected quantity, in addition to the name of the molecule, to the clipboard. (For molecule collections, quantities for all members together with member names are copied to the clipboard.) Otherwise, **Copy** copies the contents of the screen (minus the background) to the clipboard.

Select **Close All** from the **File** menu to remove all the molecules from the screen.

You are now ready to proceed. To bring all the models required for a particular problem onto the screen, *double click* on the problem title.

IV. Molecular Modeling Problems

A series of molecular modeling problems keyed to Wade's **Organic Chemistry** follows. These have been drawn from a larger compendium "The Molecular Modeling Workbook for Organic Chemistry."[1] Each problem refers to one or more pre-built models contained on the accompanying CD-ROM, and accessed via SPARTANView. "Normal" molecules as well as a variety of reactive intermediates are represented, as are reaction transition states. While most of the models are "static," some portray molecular motion involved in conformation change or chemical reaction. All of the models portray molecular structure. Most also contain energy, dipole moment and atomic charges, and a few models include vibrational frequencies. Graphical displays are available for some models and include total electron densities, electrostatic potential maps and selected molecular orbitals.

You will need to examine, manipulate and query the models on the CD-ROM in order to solve the problems. This is the essence of molecular modeling. The models contain much more information than you need, so don't be afraid to explore.

---

1. W.J. Hehre, A.J. Shusterman and J.E. Nelson, **The Molecular Modeling Workbook for Organic Chemistry**, Wavefunction, Inc., Irvine, CA, 1998.

Wade Section 1-6

Are All Chemical Bonds the Same?

Chemists refer to the bond in a molecule like sodium chloride as "ionic," meaning that its electron pair resides entirely on chlorine. At the other extreme is the non-polar "covalent" bond in the hydrogen molecule, where the electron pair is shared equally between the two hydrogens. Intermediate cases, such as the bond in hydrogen fluoride which is clearly "polarized" toward fluorine, are generally referred to as "polar covalent" bonds (rather than "partially ionic" bonds). Are these situations really all different or do they instead represent different degrees of the same thing?

Examine electron density surfaces for ***hydrogen***, ***lithium hydride***, ***beryllium hydride***, ***borane***, ***methane***, ***ammonia***, ***water*** and ***hydrogen fluoride***. First, focus on the shape of the surface (corresponding to the shape of the underlying electron density). For which molecule is the "size" of hydrogen the smallest? For which is it the largest? Is there a correlation between size of the density around hydrogen and the difference in electronegativities between hydrogen and the element to which it is bonded (see table below)? Explain.

Electronegativities

H 2.2	Li 1.0	Be 1.6	B 2.0
C 2.6	N 3.0	O 3.4	F 4.0

Next, examine electrostatic potential maps for the same set of compounds. Recall that negatively-charged regions are colored red, while positively-charged regions are colored blue. Focus your attention on the value of the potential around hydrogen. For which molecule is it most positive? For which is it most negative? Is there a correlation between the value of the potential and the difference in electronegativities? Plot charge on hydrogen vs. difference in electronegativities. Is there a correlation?

What electronegativity difference, large or small, creates a more polar bond? A more covalent bond?

Resonance Structures. The Sum of the Parts

While the majority of molecules may be adequately represented by a single resonance contributor, there are numerous situations where two or more contributors are needed. The simplest case is where all the contributing resonance structures are equivalent. Here, the proper description is in terms of an unweighted average.

Draw appropriate resonance contributors for benzene. Are all contributors equivalent? Measure the six carbon-carbon bond lengths in ***benzene***. Are they all the same? Are they intermediate in length between "normal" single bonds (in ***ethane***) and "normal" double bonds (in ***ethene***)? Is benzene properly described in terms of an equal weighting among its resonance contributors? Repeat your analysis with ***formate anion***, and address the same issues as above. Refer to ***methanol*** and ***formaldehyde*** as examples of molecules incorporating carbon-oxygen single and double bonds, respectively.

The situation is more complicated when the set of "reasonable" contributing structures are not all equivalent. Examine the geometry and atomic charges for ***phenoxide anion***. Do these data fit any one of the possible resonance structures (draw all reasonable possibilities), or is a combination of two or more resonance contributors necessary?

Wade Section 1-9

Localized vs. Delocalized Charge

Resonance theory provides a qualitative description of the location of excess (positive or negative) charge in a molecule. Each resonance contributor assigns charge to a particular center, and the extent to which charge is delocalized, and hence stabilized, may be judged simply by counting the number of contributing structures.

Draw all reasonable resonance contributors for both planar and perpendicular conformers of benzyl cation. Identify the site(s) of the positive charge in each. Which cation would you expect to be more stable? Which is the more stable? Compare energies of ***planar*** and ***perpendicular*** conformers of ***benzyl cation***.

Electrostatic potential maps provide a measure of the charge distribution in carbocations. Localized ions will show areas of high positive potential (large positive charge), while the potential in delocalized ions will be more uniform. Display electrostatic potential maps for both planar and perpendicular conformers of benzyl cation. Recall that the most positively charged regions will be colored blue. Is the charge in the lower-energy conformer more or less delocalized than that in the higher-energy conformer?

Acid-Base Properties and Partial Charge

Acids are defined as proton (H^+) donors.

$$H{-}A \rightleftharpoons H^+ + A^-$$

The HA bonds in stronger acids are polarized $H^{\delta+}{-}A^{\delta-}$ so that H is already "proton-like." Consequently, an acid's proton-donating ability ("acid strength") is usually correlated with the partial charge on hydrogen. This can be obtained from an electrostatic potential map.

Compare atomic charges and electrostatic potential maps for ***methane***, ***ammonia***, ***water*** and ***hydrogen fluoride***. Recall that electron-rich sites will be colored red while electron-poor regions will be colored blue. Which molecule contains the most electron-poor hydrogen (largest δ+)? Which molecule contains the least electron-poor hydrogen (smallest δ+)? What relationship, if any, exists between the atomic charge on hydrogen and the electronegativity of the atom bonded to hydrogen (see table below)?

Electronegativities

C 2.6	N 3.0	O 3.4	F 4.0

What relationship, if any, exists between atomic charge and experimental pK_a (see table below)?

	pK_a
CH_4	50 (est.)
NH_3	36 (est.)
H_2O	15.7
HF	3.2

Wade Section 1-13C

Acid-Base Properties and Charge Delocalization

Electron delocalization in an acid, or its conjugate base, can have a large impact on both stability and reactivity. Consider the following acids and their conjugate bases.

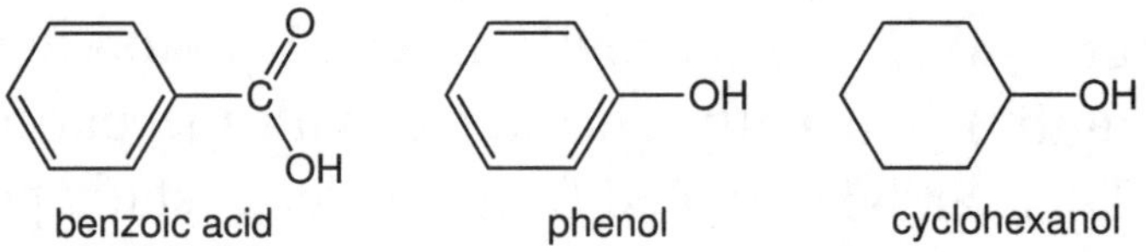

Cleavage of the OH bond gives oxygen a negative charge. However, electron delocalization may spread this charge over several atoms and stabilize the ions to varying degrees.

Compare atomic charges and electrostatic potential maps for ***benzoate anion***, ***phenoxide anion*** and ***cyclohexanoxide anion***. Recall that the most negatively-charged regions will be colored red. Which ion concentrates the most negative charge on a single atom? Which ion spreads the charge around most effectively? Which ions seem to spread charge into the ring? Is the phenyl ring or the cyclohexane ring better able to delocalize charge? Draw whatever Lewis structures are needed to describe each ion's charge distribution.

Obtain energies for each ion and for their corresponding precursors (***benzoic acid***, ***phenol*** and ***cyclohexanol***). Use this information to calculate the energy for each of the above deprotonation reactions. (The energy of proton is 0 au.) Is the trend consistent with the experimental pK_a data (see table below)?

	pK_a
benzoic acid	4.2
phenol	9.9
cyclohexanol	18 (est.)

Does deprotonation energy parallel charge delocalization in these systems? Explain how electron delocalization affects the reactivity of these acids.

Liquid Water

Water boils at a much higher temperature than would be expected based solely on its molecular weight. The reason is that liquid water exhibits a highly structured network of hydrogen bonds.

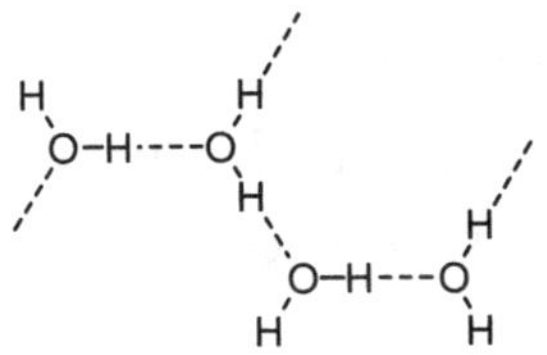

Liquid water is a sample of 36 water molecules in one of a nearly infinite number of possible ordered arrangements. Identify at least ten hydrogen bonds between pairs of water molecules. Would you characterize most of them as "linear" or as "bifurcated"?

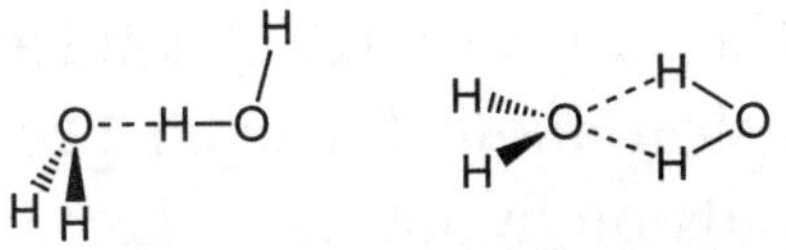

linear bifurcated

Measure at least five hydrogen-bond lengths (O–H---O) in the sample. What is the range of distances in your sample? Is the average hydrogen-bond length shorter, longer or about the same as the sum of the van der Waals radii for hydrogen and oxygen (see table below)?

van der Waals radii (Å)	
H 1.2	O 1.4

Display liquid water as a space-filling model. Are the atoms involved in hydrogen bonds "just touching" (distances ~ sum of van der Waals radii) or do they interpenetrate (distance < sum of van der Waals radii)?

Eclipsed vs. Staggered Tetrahedral Carbons

It has been observed that ethane prefers a staggered conformation. Unfortunately, experiments do not tell us why this preference exists.

Step through the sequence of structures depicting bond rotation in ***ethane***. Plot energy vs. HCCH torsion angle. Do the minima correspond to staggered structures? Do the maxima correspond to eclipsed structures?

Nonbonded atoms repel each other at short distances. This results from the fact that their electron clouds are forced so close together as to occupy, in part, the same region of space. One explanation for ethane's behavior is that staggered structures minimize "steric repulsion" between hydrogens on different carbons. Steric repulsion might be detected by its effect on HCC bond angles. Plot HCC bond angle vs. HCCH torsion angle for each geometry. Do these data support the steric repulsion hypothesis? Explain.

Estimate the "cost" of nonbonded HH repulsion as a function of distance by plotting energy (vertical axis) vs. HH separation (horizontal axis) for ***methane+methane*** (two methanes approaching each other with CH bonds "head on"). Next, measure the distance between the nearest hydrogens in eclipsed ethane. What is the HH repulsion energy in the methane dimer at this distance? Multiplied by three, does this approximate the rotation barrier in ethane?

Most molecules can avoid high-energy, sterically-hindered structures by distorting their geometries in some way. Compare the CC and CH bond distances, and the HCC bond angle, of fully staggered and eclipsed ethane. Which, if any, of these parameters undergoes distortion in order to relieve steric repulsion in the eclipsed molecule? Explain.

Steric Control of Alkane Conformation

Alkanes prefer structures that stagger bonds on adjacent tetrahedral carbons. However, steric repulsion can affect the relative energies of staggered conformations.

Step through the sequence of structures depicting bond rotation about the central carbon-carbon bond in ***n-butane***. Plot energy vs. CCCC torsion angle. How many minima and maxima are there on the graph? Draw Newman projections that show each of these conformations (draw your projection so that you are looking down the central CC bond). Label each staggered conformation as either *anti* (CCCC angle = 180°) or *gauche* (CCCC angle ≈ 60°). Do not label the maximum energy conformations.

Which conformation is more stable, *anti* or *gauche*? What evidence is there for steric repulsion between methyl groups in the *gauche* conformation? (Hint: Look for distortions in the CCC bond angles as a function of conformation.)

Examine the maximum energy conformations. What evidence is there for steric repulsion between methyl groups in the conformation that eclipses these groups?

Mechanism of Ring Inversion

Ring inversion, leading to interconversion of different ring conformers, is typically as facile a process as single-bond rotation. Particularly important are six-membered rings, where interconversion leads to interchange of axial and equatorial positions.

One after the other, step through (or animate) the sequence of structures depicting ring inversion in ***cyclohexane***, ***methylcyclohexane*** and ***trans-1,2-dimethylcyclohexane***. Is the overall "motion" involved in the ring inversion similar in all three? Describe any differences.

For each molecule, plot energy vs. "frame number." Identify all minimum and maximum energy structures. Identify all of the molecular structure(s) that might be observed experimentally (sketch structure and give frame number). If more than one structure might be observed, then calculate the composition of an equilibrium mixture of the various observable structures at 298 K (use the equation below).

$$\frac{N_1}{N_2} = e^{-1060(E_1-E_2)}$$

N_i is the number of molecules of conformer *i*
E_i is the energy of conformer *i* (in au)

Are your three energy diagrams qualitatively similar, or are there significant differences? Elaborate.

For each molecule, calculate the overall energy barrier for ring inversion in each direction. Use this barrier to calculate the half-life ($\tau_{½}$) of an individual molecule at 298 K (use the equation below).

$$\tau_{1/2} = \frac{0.69}{k}$$

$$k = 6.2x10^{12}e^{-1060\Delta E^{\ddagger}}$$

$\Delta E^{\ddagger}$ is the energy barrier (in au)

Which molecule inverts most rapidly? Most slowly? Why? (Hint: What geometrical changes are required for inversion?)

Steric Control of Ring Conformation

Alkylcyclohexanes exist as a mixture of chair conformers.

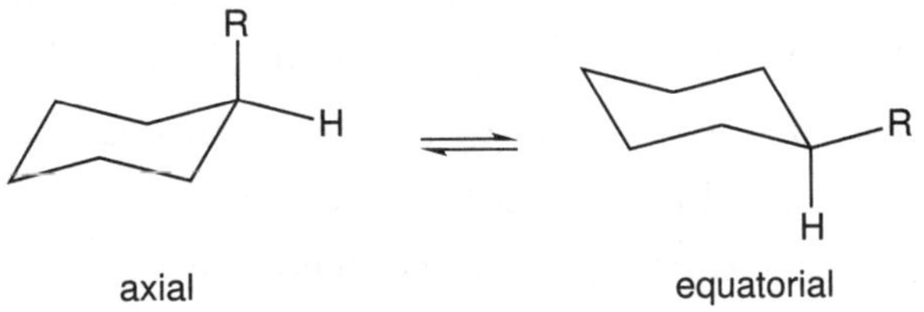

Compare energies for equatorial and axial chair conformers for ***methylcyclohexane***, R = Me, and ***tert-butylcyclohexane***, R = CMe_3. Which is more stable in each molecule? Use the equation below to calculate the ratio of major to minor conformers for each at 298 K.

$$\frac{N_1}{N_2} = e^{-1060(E_1-E_2)}$$

N_i is the number of molecules of conformer *i*
E_i is the energy of conformer *i* (in au)

Which molecule shows a larger preference? Why? (Hint: Compare nonbonded interactions and/or geometrical distortions in the higher-energy conformers that are absent in the lower-energy conformers.)

Which is more stable, the equatorial or axial chair conformer of ***i-propylcyclohexane***, R = $CHMe_2$? Calculate the ratio of major to minor conformers at 298 K. Is it more like that found for *tert*-butylcyclohexane or for methylcyclohexane? Why?

Examine the two chair conformers of ***menthol***, and label each substituent in each conformer as equatorial or axial.

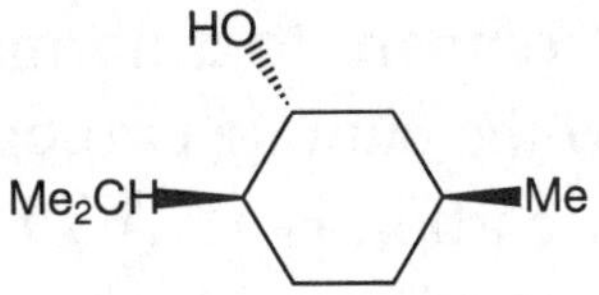

menthol

Obtain the energies of the two conformers. Which conformer is preferred? Why?

What Do Transition States Look Like?

Covalent bonds have characteristic distances depending on bond type. Carbon-carbon single and double bond lengths are around 1.54Å and 1.32Å, respectively, while partial double bond distances, e.g., in benzene, are about 1.40Å. Transition states, molecules in which bonds are being made (or broken), necessarily contain partial bonds. There are no experimental data to tell us how long these bonds are, or whether partial bonds even have characteristic distances.

Examine transition-state structures and bond density surfaces for the ***Diels-Alder***, ***ene*** and ***Cope*** reactions.

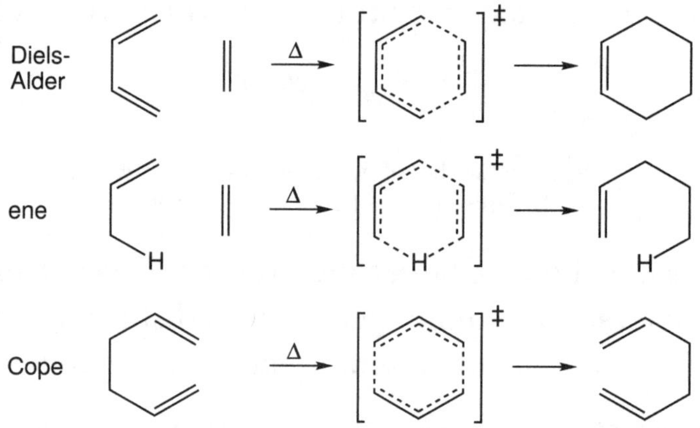

Draw the two resonance contributors that are needed to describe each transition state. Identify all partial carbon-carbon double bonds (┄┄) and measure their distances. Are these values like that found in benzene, or do transition states have their own characteristic partial double bond distance? Identify all partial single CC bonds (······) and obtain their distances. Is there a characteristic partial single bond distance? How does it compare to a normal single bond distance? How does it compare to the sum of two carbon atomic radii? (The van der Waals radius of carbon is 1.7Å.) Do bond density surfaces show a significant concentration of electrons for atoms connected by partial single bonds? Repeat your analysis for the partial CH bonds in the ene transition state. (A typical covalent CH bond distance is 1.10Å and the van der Waals radius of hydrogen is 1.2Å.)

Electronic Structure of Transition States

Chemists use curved arrows to show the electronic changes that occur during a chemical reaction. For example, the arrows describing the S_N2 reaction below show formation of a CC bond and loss of a CI bond.

However, arrows do not tell us the actual geometry or electron distribution found in the transition state. Examine the geometries of the reactants (***cyanide anion*** and ***methyl iodide***) and products (***acetonitrile***), as well as the transition state for the nucleophilic displacement (***cyanide+methyl iodide***). Obtain distances for all of the bonds in each (include the CC and CI "bonds" of the transition state). Which bonds show significant changes in distance (> 0.1Å) from reactants and which do not? Do these changes signify bond forming or bond breaking? Based on these data, draw a molecular structure for the transition state using solid lines for normal bonds and dashed lines for partially made/broken bonds.

Examine atomic charges and the electrostatic potential map for the transition state for S_N2 reaction of cyanide with methyl iodide. Recall that the most negatively charged regions will be colored red. Which atoms appear to be most electron rich? Is the negative charge concentrated on a single atom in the transition state or delocalized? Add this charge information (either "–" or "δ-") to the molecular structure for the transition state which you drew previously.

Does your transition state drawing look more like a single Lewis structure or a resonance hybrid? If the latter, what resonance contributors must you combine to generate all of the features of this hybrid?

Hyperconjugation

Resonance theory suggests that the positive charge in *tert*-butyl cation is dispersed onto the hydrogens.

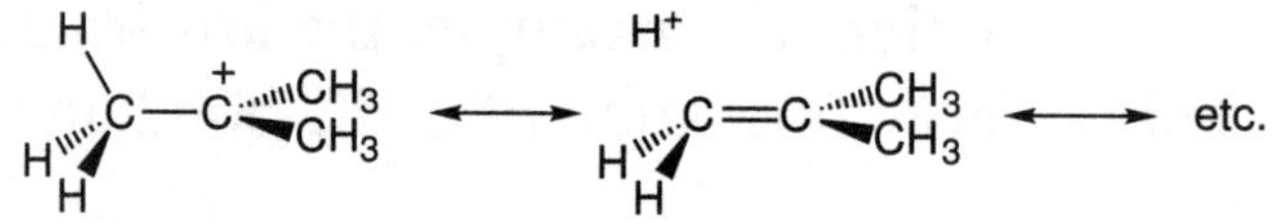

Hyperconjugation, as it is termed, implies that the electron pair associated with an out-of-plane CH bond is donated into the empty p orbital at the carbocation center.

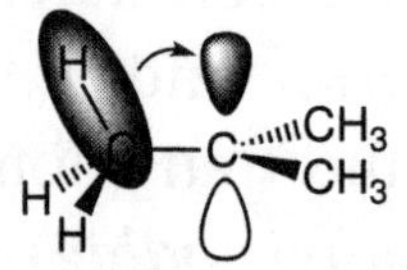

How does hyperconjugation affect the geometries of carbocations? Are the CC bonds in ***tert-butyl cation*** shorter, longer, or about the same as those in ***i-butane***? Are all CH bonds in *tert*-butyl cation the same length? Explain. Examine atomic charges and the electrostatic potential map for *tert*-butyl cation. Recall that the most positively-charged regions will be colored blue. Relate any differences between in-plane and out-of-plane hydrogens both to the resonance picture and to differences in CH bond lengths.

Step through the sequence of structures depicting rotation about the C_{Et} - C^+ bond in ***2-methyl-2-butyl cation***. Plot energy vs. CCCC dihedral angle. Is the conformation with the ethyl group in plane or perpendicular to the plane preferred?

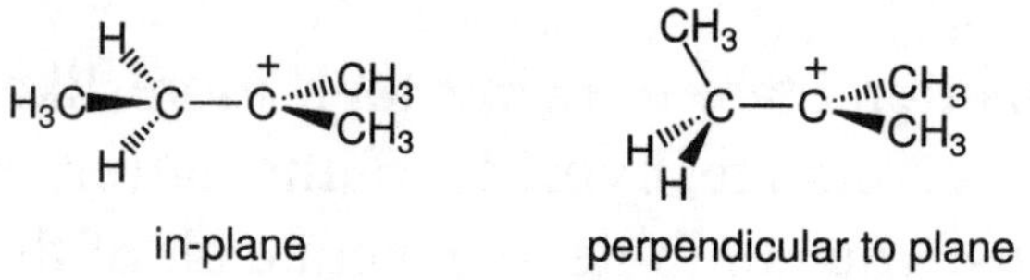

Does the C_3C_4 distance change with conformation? Do you see evidence for CC hyperconjugation? What does your result tell you about the relative importance of CH and CC hyperconjugation? Explain.

Structure of Free Radicals

What is the preferred geometry about the radical center in free radicals? Carbocation centers are characterized by a vacant orbital and are known to be planar, while carbanion centers incorporate a nonbonded electron pair and are typically pyramidal.

First examine the geometry of ***methyl radical***. Is it planar or puckered? Examine the geometries of ***tert-butyl radical***, ***trifluoromethyl radical***, ***trichloromethyl radical*** and ***tricyanomethyl radical***. Classify each of the substituents (methyl, fluoro, chloro and cyano) as a π- electron donor or as a π-electron acceptor (relative to hydrogen). Does replacement of the hydrogens by π-donor groups make the radical center more or less puckered? Does replacement by π-acceptor groups make the radical center more or less puckered? Justify your observations.

Display spin density surfaces for all radicals. These depict the location of the unpaired electron. For which radical is the unpaired electron least delocalized from the radical center? For which is it the most delocalized? Is there any relationship between degree of puckering of the radical center and extent of spin delocalization?

Onto which atoms (carbon, nitrogen or both) is the unpaired electron in tricyanomethyl radical delocalized? Rationalize your result by drawing resonance contributors.

Singlet and Triplet Methylene

Methylene (CH_2) has six valence electrons. Four are needed for the two CH bonds. Possibilities for the other two include:

The first two arrangements are singlet states (all electrons are paired), while the last is a triplet state (with two unpaired electrons). Experimentally, the ground state of methylene is a triplet, although much of methylene's chemistry (and that of substituted methylenes) is due to the singlet state.

Examine the highest-occupied molecular orbital (HOMO) of ***singlet methylene***. This shows location of the molecule's highest energy pair of electrons. Where is the pair of electrons, in-plane or perpendicular to the plane? Next, examine the electrostatic potential map. This shows electron-rich regions in red and electron-poor regions in blue. Where is the molecule most electron rich, in the σ or the π system? Where is the most electron poor? Next, display the corresponding map for ***triplet methylene***. Which molecule would you expect to be the better nucleophile? The better electrophile? Explain. Experimentally, one state of methylene shows both "electrophilic" and "nucleophilic" chemistry, while the other state exhibits chemistry typical of radicals. Which state does which? Elaborate.

Which is lower in energy, singlet or triplet methylene? What effect do substituents have on altering the singlet-triplet energy difference in methylene? One after the other, compare energies for ***singlet*** and ***triplet difluoromethylene*** and ***singlet*** and ***triplet dicyanomethylene***, and identify the ground state for each. Does fluorine substitution favor the singlet or triplet state? Does cyano substitution favor the singlet or triplet state? Rationalize your observations. (Hint: Compare geometries among the three methylenes for both singlet and triplet states.)

Enantiomers

Most of the molecules we take into our bodies, whether in food or in medicine, are chiral. As a rule, different enantiomers of these molecules have different biological behavior.

Each of the molecules below (carvone, ibuprofen and limonene) incorporates a single chiral center. Identify it and draw both R and S forms of each.

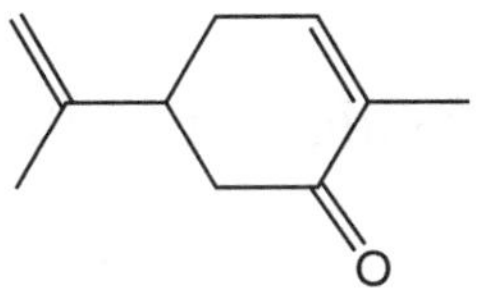

carvone

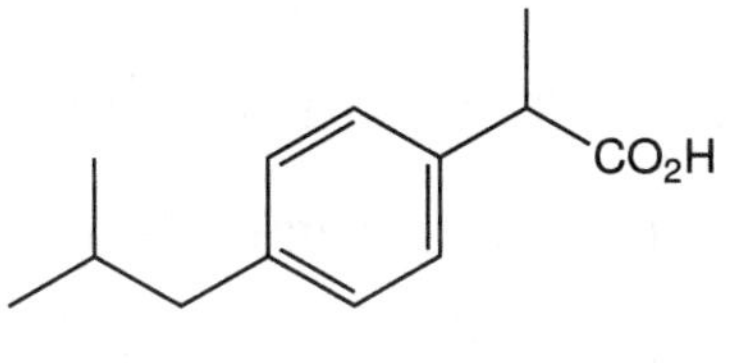

ibuprofen

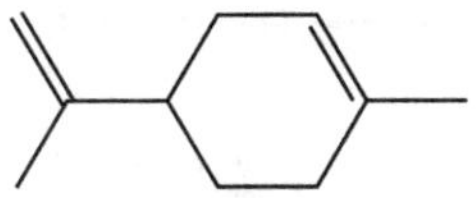

limonene

Examine actual three-dimensional structures for the two enantiomers of ***carvone***. One occurs naturally in caraway while the other is found in spearmint oil, and are responsible for the characteristic odors of these materials. Determine which form, R or S, is responsible for which odor.

Ibuprofen is an analgesic sold under various names, including Advil, Motrin, and Nuprin. The material is sold as a racemic mixture, but only one enantiomer acts as an analgesic. The other enantiomer is inactive. Assign R or S forms to the two enantiomers of ***ibuprofen***.

The two enantiomers of limonene have completely different tastes. One has the taste of lemon (as the name implies) and the other of orange. Assign R or S forms to the two enantiomers of ***limonene***.

Are the energies and dipole moments for the two enantiomers of carvone (ibuprofen and limonene) the same or are they different? Explain your result.

Chromatography and Molecular Polarity

Chromatography is an important practical methodology for separating mixtures of organic compounds. While there are many chromatographic techniques, all basically involve passing the mixture of compounds to be separated over an immobile support contained in a column (the "stationary phase"). Molecules that "stick" strongly to the stationary phase pass more slowly through the column than molecules that stick less strongly.

Oxidation of sulfides results both in sulfoxides and sulfones, as well as starting material.

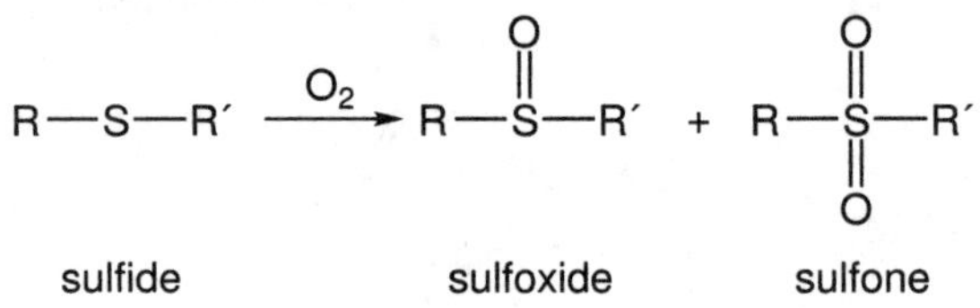

These can usually be easily separated by thin layer chromatography (TLC) on silica gel.

Measure dipole moments and atomic charges, and display and compare electrostatic potential maps for ***methyl cyclohexyl sulfide***, ***sulfoxide*** and ***sulfone***. Which molecule has the largest dipole moment? The smallest? Focusing only on the functional groups, which atoms in each are most positively charged? Most negatively charged? Recall that negatively-charged regions are colored red in the electrostatic potential maps while positively-charged regions are colored blue. Does increased oxidation lead to sulfur becoming more positively charged, more negatively charged or leave it unchanged? Explain. Overall, which molecule is most polar (positive and negative charge most widely separated) and which is least polar? Were a mixture of these molecules to be dissolved in a non-polar solvent and passed over a highly-polar stationary phase, which isomer would you expect to elute first and which would you expect to elute last? Explain your reasoning.

S_N2 and Proton-Transfer Reactions

S_N2 reactions can be thought of as "alkyl transfer" reactions, and "S_N2 characteristics" can be anticipated by examining analogous proton transfer reactions.

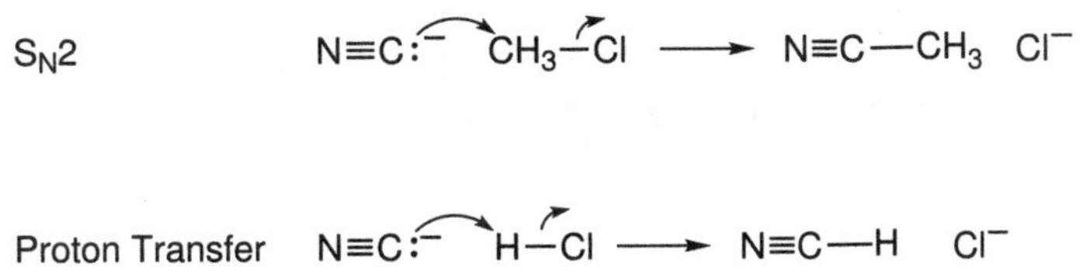

One after the other, step through (or animate) the sequence of structures depicting the ***Sn2*** and ***proton transfer*** reactions shown above. Compare the two. From what direction does cyanide approach the hydrogen in HCl? From the same side as Cl ("frontside"), or from the other side ("backside")? Does the S_N2 reaction follow a similar trajectory?

Describe what happens to the negative charge during the course of the proton transfer reaction. Plot charge on "CN^-" (sum of the charges on C and N) vs. number of the frame in the sequence. Also on the same graph, plot the charge on hydrogen (vertical axis). Is a "proton" generated during the course of reaction, or is charge buildup limited as the hydrogen is passed between chlorine and carbon?

Repeat the analysis for the S_N2 reaction. Is there significant buildup of positive charge on "CH_3"?

Steric Hindrance of S_N2 Reactions

S_N2 reactions proceed through transition states in which the central carbon has five neighbors instead of the usual four, e.g., for reaction of bromide and methyl bromide.

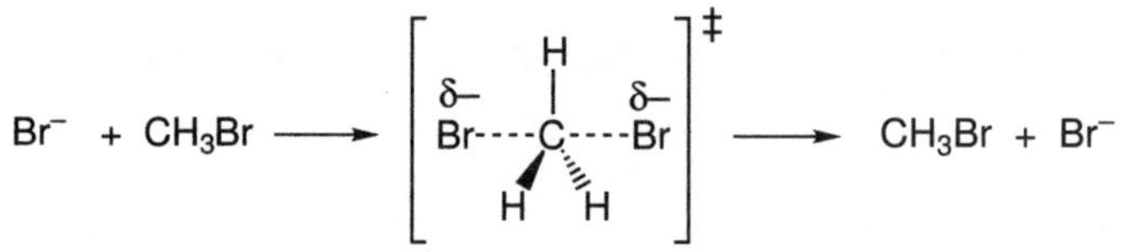

Does this imply that steric effects will be "amplified" in the transition state, and that the rates of S_N2 reactions will decrease with increased substitution at carbon?

Calculate activation barriers for bromide addition to ***methyl bromide*** and ***tert-butyl bromide*** using energies for S_N2 transition states (***bromide+methyl bromide*** and ***bromide+tert-butyl bromide***). The energy for Br^- is -2560.2998 au. Which reaction is faster?

Next, examine the S_N2 transition states as space-filling models. Are you able to identify unfavorable nonbonded (steric) interactions that are not present in the reactants? If so, which S_N2 reaction is likely to be more affected by steric interactions? Rationalize your observations. Hint: Compare CBr bond distances in the S_N2 transition states. How do these change with increased substitution at carbon? What effect, if any, does this have on crowding?

What other factors might be responsible for differences in activation energies? Compare atomic charges and electrostatic potential maps for the S_N2 transition states. Recall, that electrostatic potential maps depict the most negatively-charged regions in red. Does the increase in steric crowding lead to enhanced or diminished charge delocalization? Explain. How, if at all, would this be expected to affect the energy barrier? Why?

What is the origin of the change in rate of S_N2 reactions with change in substitution at carbon?

Stereochemistry of S_N2 Reactions

S_N2 reactions proceed with inversion at the electrophilic carbon. This suggests that the nucleophile attacks from the "backside" of carbon, i.e., the side of carbon furthest away from the leaving group.

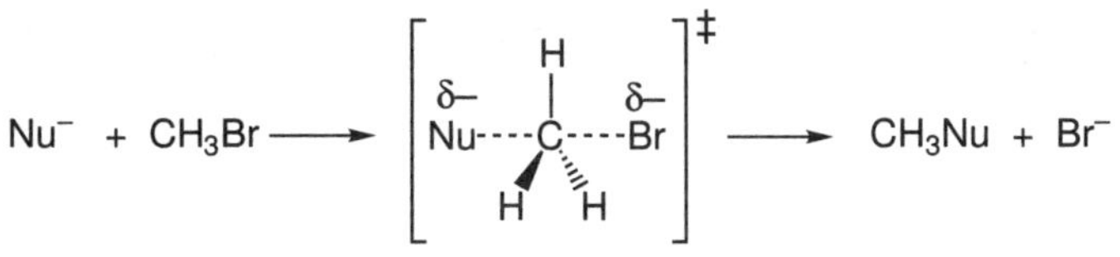

Backside attack may be favored for electrostatic reasons. Examine electrostatic potential maps for ***bromide+methyl bromide frontside attack*** and ***bromide+methyl bromide backside attack***, transition states involving "frontside" and "backside" attack of Br^- (the nucleophile) onto CH_3Br, respectively. Recall that the most negatively-charged regions will be colored red and less negatively-charged regions will be colored blue. Which atoms in the transition states are most electron-rich? Which trajectory better minimizes electrostatic repulsion?

Backside attack may be favored in order to facilitate transfer of nonbonding electrons from the nucleophile into the electrophile's lowest-unoccupied molecular orbital (LUMO). Efficient electron transfer requires maximal overlap of the LUMO and the donor orbital (usually a nonbonded electron pair on the nucleophile). Examine the LUMO of ***methyl bromide***. How would a nucleophile have to approach in order to obtain the best overlap? Is your answer more consistent with preferential "backside" or "frontside" attack?

Electron transfer into the LUMO might also cause bonding changes. What are the CBr bonding characteristics of the LUMO in methyl bromide? Is it "bonding" (one surface extends over the bond) or "antibonding" (two surfaces meet in middle of the bond)? How would electron transfer from a nucleophile affect the CBr bond length?

Wade Section 6-13A

Stability of Carbocation Intermediates

tert-Butyl bromide reacts rapidly with ethanol to give ethyl *tert*-butyl ether. This does not involve S_N2 displacement, but rather the reaction proceeds via a two-step "S_N1" mechanism and a carbocation intermediate.

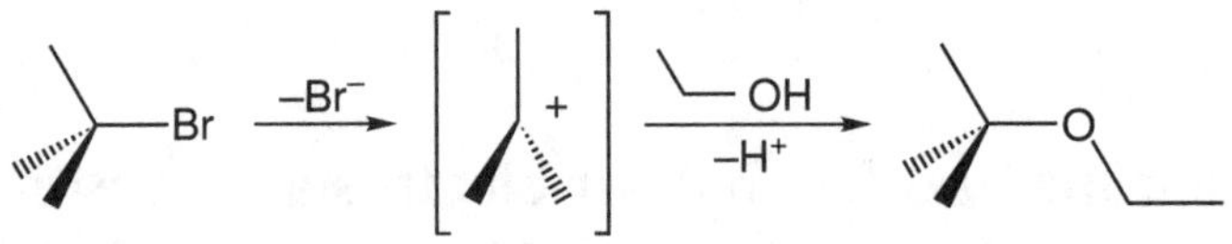

S_N1 reactivity follows the order: 3°>2°>1°. Is this the ordering of stabilities of the carbocation intermediates?

Display and examine electrostatic potential maps for ***ethyl cation***, ***i-propyl cation*** and ***tert-butyl cation***. Which cation shows the greatest localization of positive charge (blue color)? If you find that the methyl groups delocalize the positive charge, where does the charge go? Write resonance contributors for the three cations to rationalize your conclusion. (Note: You may need to draw resonance contributors that contain a CC double bond and are missing a CH bond.)

Examine and compare atomic charges and electrostatic potential maps for *tert*-butyl cation and ***3-ethyl-3-pentyl cation***. Are the methyl groups in *tert*-butyl cation and the methylene and methyl groups in 3-ethyl-3-pentyl cation positively charged? Are the ethyl groups in 3-ethyl-3-pentyl cation more effective, less effective or about as effective as the methyl groups in *tert*-butyl cation in delocalizing positive charge? Are the methylene groups in 3-ethyl-3-pentyl cation more positively charged than the methyl groups? Do your results support or refute the usual inductive picture?

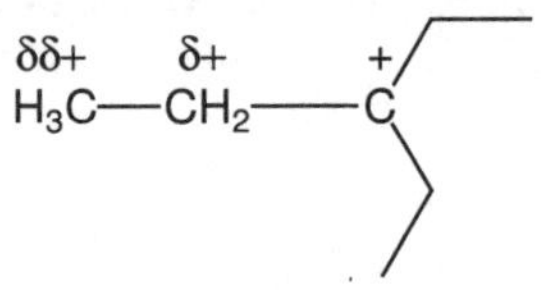

Phenyl vs. Benzyl Cation

Benzene rings have a dramatic effect on S_N1 reaction rates. This depends on the position of the ring relative to the leaving group. Consider the following reactions.

$$Me_3CCl \longrightarrow Me_3C^+ + Cl^- \quad (1)$$

$$PhC(Me)_2Cl \longrightarrow PhCMe_2^+ + Cl^- \quad (2)$$

$$PhCl \longrightarrow Ph^+ + Cl \quad (3)$$

Calculate energies for reaction of ***tert-butyl chloride*** (to ***tert-butyl cation***), ***2-phenyl-2-propyl chloride*** (to ***2-phenyl-2-propyl cation***) and ***chlorobenzene*** (to ***phenyl cation***). (The energy of chloride is -457.4441 au.) Assuming that reaction (1) is "normal," what is the effect of a benzene ring? Does it facilitate or hinder loss of chloride?

Compare CCl distances in the reactants. If shorter bonds are stronger, which reactants will show unusual reactivity, and in what direction? Compare the cation structures with each other and with the reactants. If the structural changes in reaction (1) are "normal," then what unusual changes, if any, occur in reactions (2) and (3)?

Compare atomic charges and electrostatic potential maps for the three cations. Recall, that the most positively-charged regions will be colored blue. For each, is the charge localized or delocalized? Draw all of the resonance contributors needed for a complete description of each cation. Assign the hybridization of the "C^+" atom, and describe how each orbital on this atom is utilized (σ bond, π bond, empty). How do you explain the benzene ring effects that you observe?

Based on your results, what would you expect is the product of the following reaction?

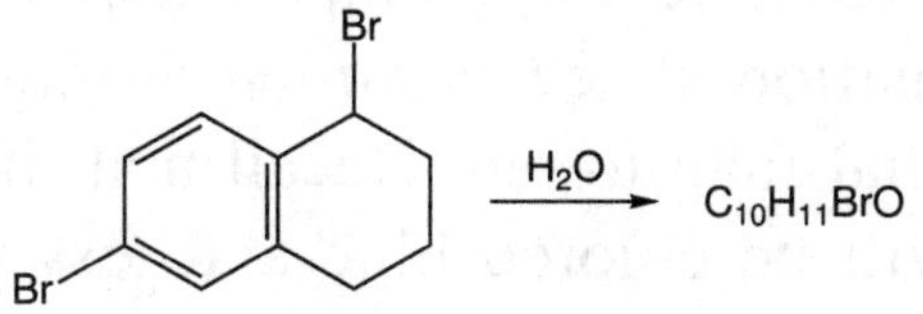

Skeletal Rearrangements of Carbocation Intermediates

Carbocations initially formed upon addition of an electrophile to an alkene may be able to undergo skeletal rearrangement depending on whether or not a more stable cation exists and, if it does exist, whether or not it can be reached via a low-energy pathway. Consider addition of HBr to 3-methyl-1-butene, the product of which is 2-methyl-2-butyl bromide.

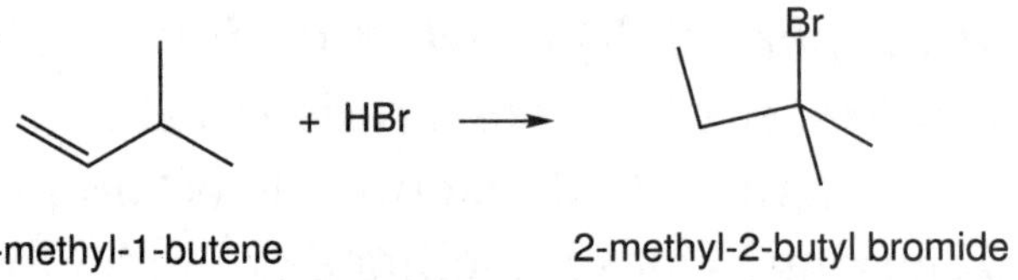

Draw Lewis structures for the possible carbocations resulting from protonation of the double bond in 3-methyl-1-butene, and decide which is favored. (Check your result using available energy data for C_5H_{11} ***carbocations***.) What would be the product of bromide addition to the more stable cation? Is this the observed product?

Draw a Lewis structure for the carbocation which would result from a 1,2-hydride shift in the more stable (initially-formed) cation. Is the rearrangement exothermic? What would be the product of bromide addition to the rearranged cation? Is this the observed product?

Examine the transition state for the ***hydride shift***. Calculate the barrier from the more stable initial carbocation. Is the process more facile than typical thermal rearrangements of neutral molecules (.05 to .08 au or approximately 30-50 kcal/mol)? Is the barrier so small (<.02 au or approximately 12 kcal/mol) that it would be impossible to stop the rearrangement even at very low temperature? Where is the positive charge in the transition state? Examine atomic charges and the electrostatic potential map to tell. Recall that the most positively-charged regions will be colored blue and less positively-charged regions colored red. Is the name "hydride shift" appropriate? If not, propose a more appropriate name.

Conformational Control of E2 Elimination

Base-promoted E2 elimination involves simultaneous loss of H^+ and X^- from neighboring carbons. This suggests that elimination in 2-methylcyclohexyl tosylate might lead to two different products. The actual situation is more complicated. One tosylate isomer gives only one of the two possible alkenes, while the other gives both.

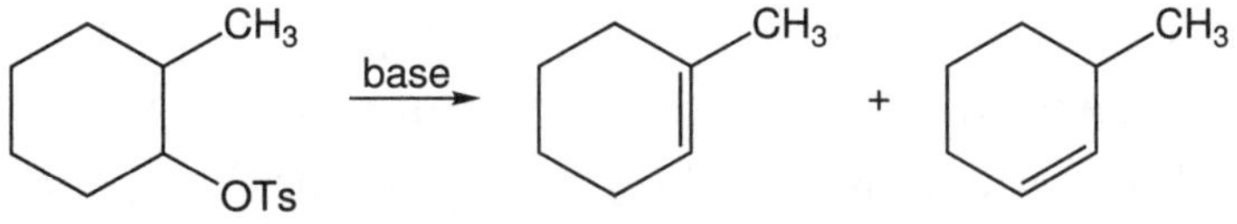

Examine all of the low-energy (within .004 au or ≈ 3 kcal/mol of the lowest-energy conformer) conformers of ***cis-2-methylcyclohexyl tosylate***. Identify every conformer that can undergo *anti* elimination of $^-$OTs and H^+, and predict the alkene that will be produced. What alkenes will be obtained from the *cis* tosylate?

Analyze the low-energy conformers of ***trans-2-methylcyclohexyl tosylate*** in the same way. What alkenes will be obtained from the *trans* tosylate?

Another interesting question concerns the rate at which each tosylate undergoes elimination. A tosylate sample contains molecules with several different conformations. The size of each conformer population depends on conformer energy, and the more reactive tosylate will probably be the one with the largest population of "reactive" conformers, i.e., molecules whose geometries allow *anti* elimination. Which tosylate, *cis* or *trans*, will have a larger population of "reactive" conformers? Explain how you reached this conclusion.

cis-trans Isomerization

Two different isomers of 2-butene exist. Which isomer, ***cis-2-butene*** or ***trans-2-butene***, is lower in energy? Compare space-filling models to see if one molecule is more crowded than the other, and dipole moments to see if one is more polar than the other. What do you suspect is the origin of the thermodynamic preference?

Calculate the activation barrier for *cis-trans* isomerization as the difference in energies between *cis*-2-butene and the ***transition state*** for *cis-trans* isomerization. Is it smaller, larger, or about the same as the energy required for *gauche-trans* isomerization in *n*-butane (*see Page 35*)? What is the origin of the barrier? Hint: Examine the central carbon-carbon bond in the transition state for *cis-trans* isomerization in 2-butene. Is it still a double bond (as in either *cis* or *trans*-2-butene) or has it lengthened to the value in ***n-butane***? Also compare the highest-occupied molecular orbital (HOMO) of the transition state with that for either *cis* or *trans*-2-butene. What, if anything, does this tell you about the integrity of the double bond in the transition state?

Cis-trans isomerization of retinal is important in the chemistry of vision.

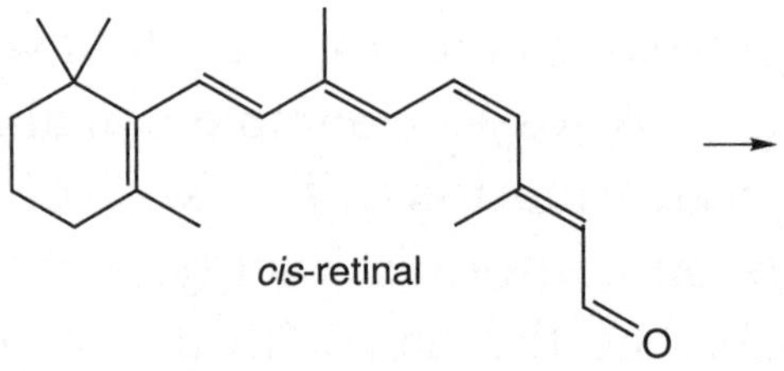

cis-retinal

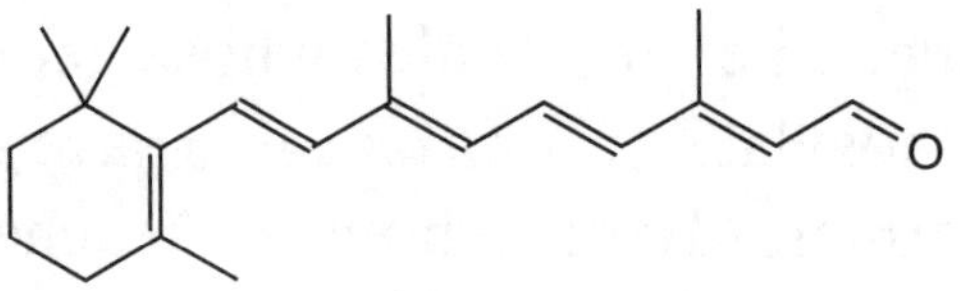

trans-retinal

Compare energies for ***cis*** and ***trans-retinal***. Is isomerization endothermic or exothermic? Why?

Hydroboration of Alkenes

Conversion of alkenes to alcohols by hydroboration is a synthetically-valuable reaction as it leads to the *anti*-Markovnikov product.

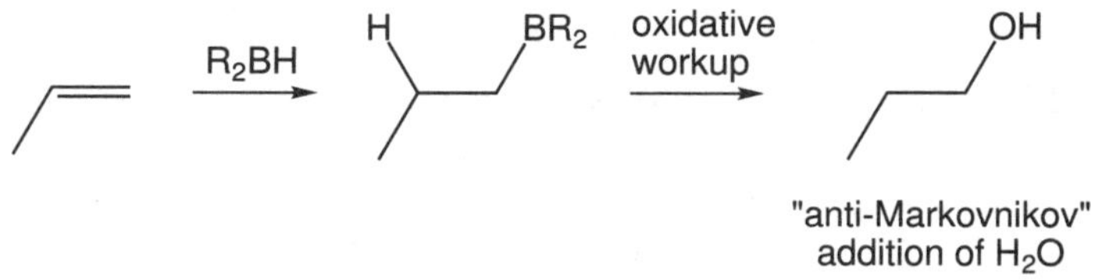

Dimethylborane+propene C1 depicts the transition state for addition of dimethylborane onto the terminal alkene carbon of propene. Examine and describe the vibration with the "imaginary" frequency. This corresponds to motion along the reaction coordinate. Which bonds stretch and compress the most? What simultaneous changes in bonding are implied by these motions? Simultaneously display the highest-occupied molecular orbital (HOMO) of ***propene*** and the lowest-unoccupied molecular orbital (LUMO) of ***dimethylborane***. Is the overall geometry of the transition state consistent with constructive overlap between the two? Explain.

Obtain the energies of propene, dimethylborane, and ***1-propyldimethyl borane***, and calculate ΔH_{rxn} for dimethylborane addition. Is this reaction exothermic or endothermic? Use this result and the Hammond Postulate to predict whether the transition state will be more "reactant like" or more "product like." Compare the geometry of the transition state to that of the reactants and products. Does the Hammond Postulate correctly anticipate the structure of the transition state? Explain.

Dimethylborane+propene C2 and ***2-propyldimethyl borane*** depict the regioisomeric transition state and addition product. Calculate the energies of these species relative to those of the alternative transition state and product. Given these energy differences, and the experimental observation that this addition is almost completely selective for the anti-Markovnikov product, does it appear that this reaction is under kinetic or thermodynamic control? Explain.

Stereochemistry of Alkene Hydrogenation

Alkene hydrogenation occurs on the surface of metal particles which act as a catalyst for the reaction. This usually means that both hydrogens are added to the same face of the alkene (*syn* addition).

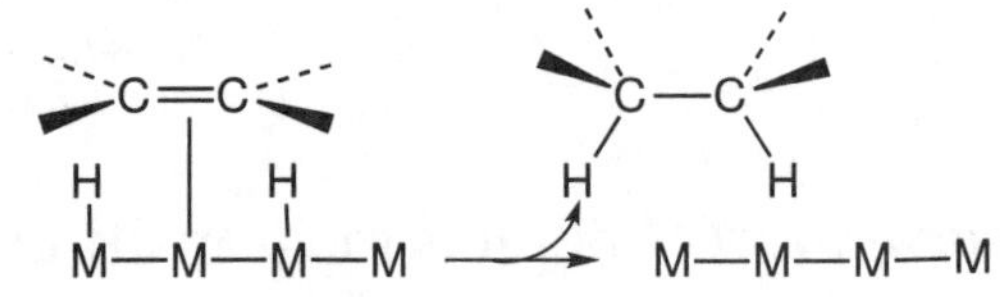

Hydrogenation of the following alkenes gives a single stereoisomer.

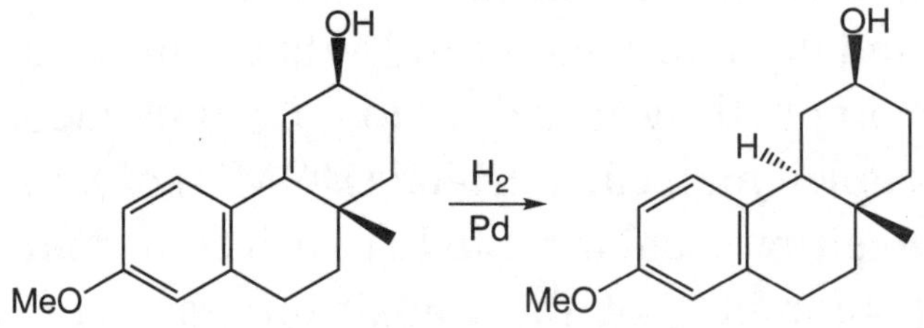

alkene A

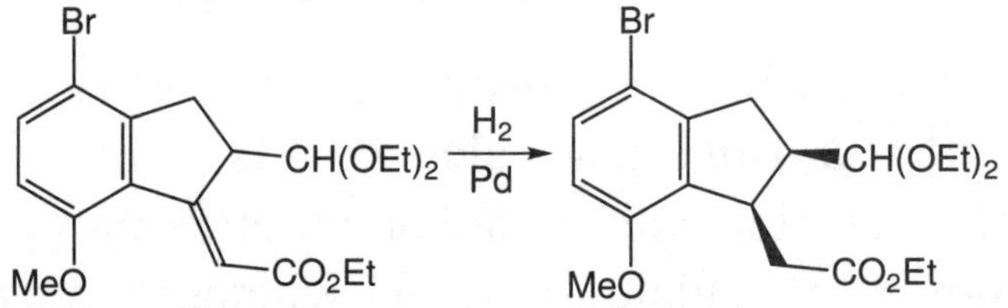

alkene B

Is the observed product for each addition also the thermodynamic product? Compare energies for ***alkene A+ H2 observed*** and ***not observed*** and ***alkene B+H2 observed*** and ***not observed***. What structural factors seem to be responsible for their relative stabilities?

Examine space-filling models and electron density surfaces for ***alkene A*** and ***alkene B***. Both portray the extent to which the different faces of the double bond are hindered. For each, which face of the double bond is less hindered? Which atoms cause steric hindrance of the alkene? Is this reaction controlled by steric hindrance? If so, explain which step(s) in the catalytic mechanism would be most affected.

Electrophilic Addition of Br_2 to Alkenes

Addition of hydrogen halides to alkenes is not stereospecific. In contrast, addition of Br_2 proceeds with *anti* stereochemistry.

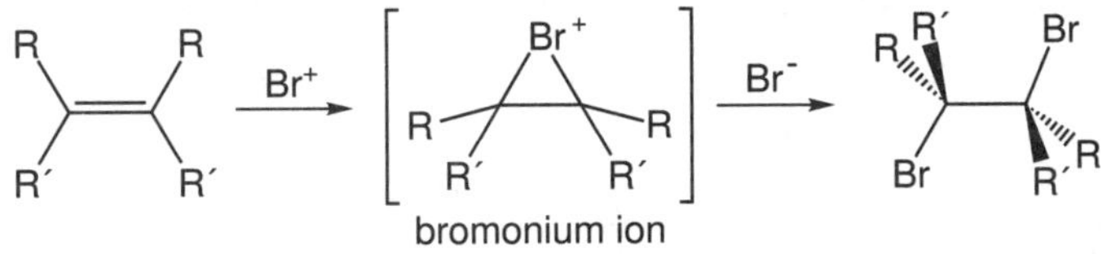

In order to explain the observed product, a cyclic "bromonium ion" intermediate has been proposed.

Examine the geometry and atomic charges of ***cyclohexylbromonium ion***, the proposed intermediate in the electrophilic bromination of cyclohexene. Measure both carbon-bromine bond distances. Are they shorter, longer or about the same as in typical alkyl bromides, e.g., ***methyl bromide***? Is the bromine centered between the two carbons? Measure the carbon-carbon bond distance. Is it consistent with a single bond or more typical of a double bond? (Compare with structures for ***cyclohexane*** and ***cyclohexene***.) Does bromide bear a full positive charge, or has some charge dispersed onto the rest of the ion? Examine atomic charges and the electrostatic potential map. Recall that the most positively-charged regions are colored blue. Draw an appropriate Lewis structure (or series of Lewis structures) to account for the geometry and charge distribution of cyclohexylbromonium ion? Is it best portrayed as a "ring" or as a weak complex between Br^+ and cyclohexene? Elaborate.

Display the lowest-unoccupied molecular orbital (LUMO) for cyclohexyl bromonium ion. This reveals the likely site for nucleophilic attack. From which side will the Br^- attack? Will this lead to formation of *cis*-1,2-dibromo-cyclohexane or *trans*-1,2-dibromocyclohexane? Is this also the thermodynamic product? Compare energies of ***cis-1,2-dibromocyclohexane*** and ***trans-1,2-dibromocyclohexane***.

Anions from Alkynes

One way to generate carbanions is to combine an acidic molecule with one equivalent of a very strong base, such as *n*-butyl lithium (*n*–BuLi). For example, reaction of the alkyne shown below with *n*-BuLi leads to a carbanion of formula $C_8H_{11}O_2^-$, which then undergoes an S_N2 reaction with *n*-propyl bromide (*n*-PrBr).

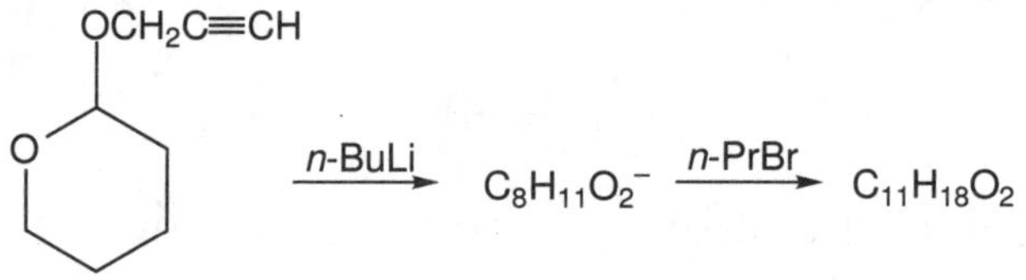

You should be able to predict the structure of the product by determining which hydrogen in the starting material is most acidic, that is, by assigning the structure of the intermediate carbanion.

First, attempt to identify the most acidic hydrogen in the starting material, based on hybridization or on the nature of neighboring atoms. Explain your rationale. Next, examine the electrostatic potential map for starting material (***alkyne***). Recall that the most positively-charged regions (the likely acidic sites) will be colored blue. Which hydrogen appears to be most electron poor? Is this the one that you predicted? What makes this hydrogen more electron poor than the others?

Note, that the $OCH_2C{\equiv}CH$ group in the alkyne starting material occupies an axial position. Why is this unexpected? Offer an explanation.

Alkyne vs. Alkene Reactivity

A simple-minded picture suggests that the CC π bonds in alkynes and alkenes ought to be similar. Are they? Consider the thermodynamics of reduction of ***phenylacetylene*** to first give ***styrene*** and then ***ethylbenzene***. (The energy of H_2 is -1.1230 au.)

$$PhC{\equiv}CH + H_2 \longrightarrow PhCH{=}CH_2 \quad (1)$$

$$PhCH{=}CH_2 + H_2 \longrightarrow PhCH_2CH_3 \quad (2)$$

Which addition is more favorable thermodynamically? Assuming that the difference is entirely due to different π-bond energies, then which contains the stronger π bond, the alkyne or the alkene? What flaws might there be in the basic assumption?

Metals that catalyze reaction (2) generally catalyze reaction (1) as well, and the rate for (1) is usually somewhat higher. Given such a catalyst, what product(s) would be obtained were phenylacetylene to be combined with one equivalent of H_2? What would happen if the catalyst accelerated reaction (2) more effectively than reaction (1)?

Thermodynamics and kinetics need not go hand in hand (*see also Page 64*). Consider all possible products resulting from addition of one equivalent of bromine to phenylacetylene (***phenylacetylene+Br2***) and to styrene (***styrene+Br2***). Calculate the heat of reaction for each addition. (The energy of Br_2 is -5120.5302 au.) Is addition to the alkyne or to the alkene more favorable?

Bromination usually follows a two-step mechanism, the rate-limiting step involving formation of an adduct with Br^+. Calculate energies for Br^+ addition to phenylacetylene and styrene, leading to ***phenylacetylene+Br+*** and ***styrene+Br+***, respectively. (The energy of Br^+ is -2559.7644 au.) Which reaction is more favorable? Is this the same preference as seen for Br_2 addition?

pK_a's of Alcohols

Alcohols are typically very weak acids with pK_a values in the range of 7 - 20 (compared with a pK_a value of 4.8 for acetic acid).

alcohols	pK_a
methanol	15.5
ethanol	15.9
2-propanol	18
2,2,2-trifluoroethanol	12.4
phenol	10.0
4-nitrophenol	7.2

Display and compare electrostatic potential maps for ***methanol***, ***ethanol***, ***2-propanol*** and ***trifluoroethanol***. Recall that the most positively-charged regions (the likely acidic sites) will be colored blue. Identify the acidic sites as those where the potential is most positive and, assuming that the more positive the potential the more acidic the site, rank the acidities of the compounds. Does increased alkyl substitution have a significant effect on acid strength? What is the effect of replacing the methyl group in ethanol by a trifluoromethyl group? Why? Do you find a correlation between the most positive value of the potential and the experimental pK_a?

Phenol has different chemical properties from those of typical alcohols. Display the electrostatic potential map for ***phenol***. Does this suggest that phenol is likely to be a stronger or weaker acid than any of the compounds discussed above? Compare the electrostatic potential map for ***4-nitrophenol*** to that for phenol. What effect does substitution by nitro have on acid strength? Explain your result by considering charge delocalization in the conjugate base. Draw all reasonable Lewis structures for phenoxide anion and for 4-nitrophenoxide anion. Which is more delocalized? Is this consistent with experimental pK_a's?

Obtain the charge on the atom for which the electrostatic potential is most positive in each of the above molecules. Plot this charge vs. experimental pK_a. Is there a correlation?

The Pinacol Rearrangement

The pinacol rearrangement is a dehydration reaction that converts a 1,2-diol into a ketone. The reaction involves two carbocation intermediates.

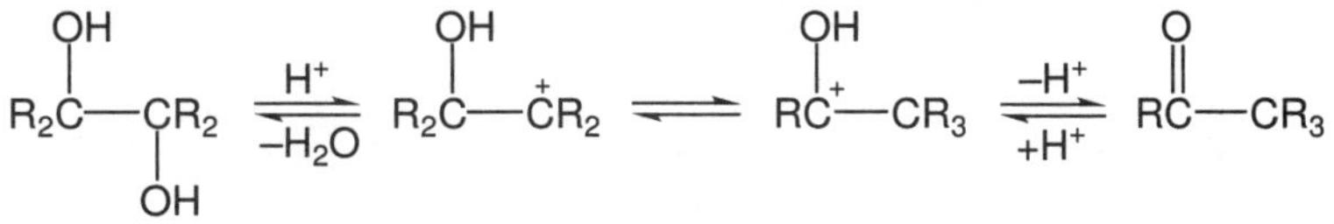

Obtain energies for the carbocation intermediates for the case of R = CH_3 (***2,3-dimethyl-3-hydroxy-2-butyl cation*** and ***3,3-dimethyl-2-hydroxy-2-butyl cation***). Is the carbocation rearrangement exothermic? Compare electrostatic potential maps for the two carbocations. Recall that the most positively-charged regions will be colored blue. Is positive charge more delocalized in the more stable cation? Why is one cation more stable than the other?

Unsymmetrical diols typically give a mixture of rearrangement products. For example, the diol shown below might give eight distinct products (counting *cis* and *trans* diastereomers as distinct products). In fact, it gives only the two shown.

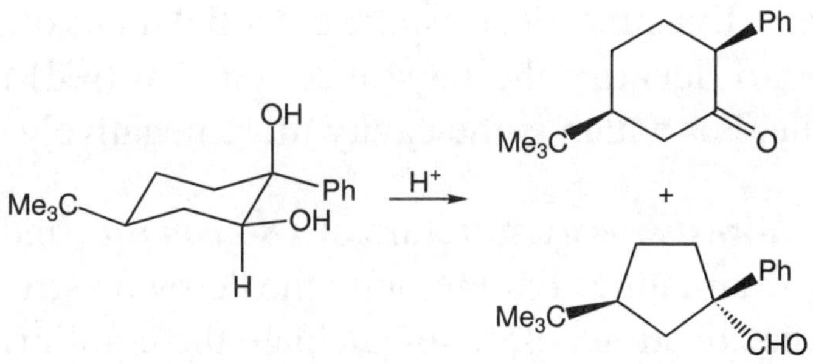

Draw the six other rearrangement products. Selectivity is controlled, in part, by the site of carbocation formation. Use the energies of ***conjugate acids*** and ***carbocations*** to calculate the energy required to form each carbocation (the energy of H_2O is -78.5860 au). Which carbocation forms more easily? Why? Is this the carbocation that leads to the observed products? Explain. Which groups migrate (and to which face of the carbocation carbon) to generate the observed products?

Crown Ethers

Crown ethers are cyclic polyethers. They may contain a cavity that can partially engulf atomic ions. For example, 18-crown-6 binds K^+ so tightly that it can extract this ion into benzene from water, driving counterions into the benzene layer.

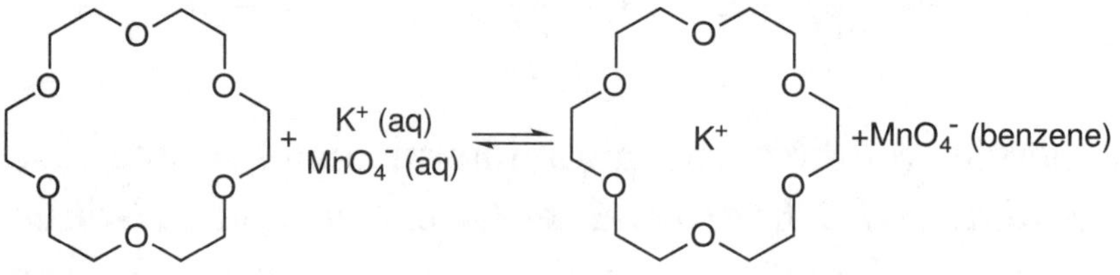

18-crown-6

Different crown ethers bind selectively to different size cations. Compare the sizes of ***lithium***, ***sodium*** and ***potassium cations*** with the size of the cavities in ***12-crown-4*** and ***18-crown-6*** (use space-filling models to estimate molecular size). Predict the relative binding ability of each crown ether for the three ions. (Assume that ion binding falls in the order: tight fit > loose fit >> too tight a fit.) Ion-crown ether interactions are electrostatic in nature. Cations will be attracted to the electron-rich environment created by the nonbonding electrons on the oxygens. Examine electrostatic potential maps of 12-crown-4 and 18-crown-6. Identify the most electron-rich (red) region(s) in each molecule. For which is the cavity more negatively charged?

Examine the ***lowest energy*** structure of ***18-crown-6***, and compare it to the "crown" structure. Explain why the "crown" structure is less stable. Use the equation below to calculate the equilibrium ratio of lowest-energy and "crown" conformers of 18-crown-6 at room temperature.

$$\frac{N_{lowest}}{N_{crown}} = e^{-1060(E_{lowest}-E_{crown})}$$

N_i is the number of molecules of conformer *i*
E_i is the energy of conformer *i* (in au)

What causes a shift in conformation in the presence of metal cations?

Resonance Control of Conformation

The following type of resonance requires π–type orbital overlap between the central carbons. This overlap can only be achieved if all four carbons and their substituents lie in the same plane.

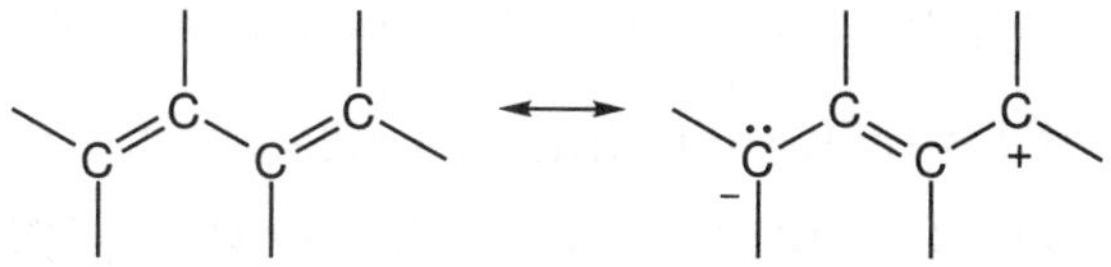

Plot the energy of ***E-1,3-pentadiene*** vs. $C_1C_2C_3C_4$ torsion angle. How many minima are there? Do they correspond to structures that offer maximum π–type orbital overlap? How can you account for differences in their energies?

What is the maximum energy structure? Does it correspond to a structure that prevents π-type orbital overlap? What is the barrier to rotation about the C_2–C_3 single bond?

Repeat this analysis for ***Z-1,3-pentadiene***.

Which isomer, E or Z, has stronger conformational preferences? Are these preferences due to resonance effects or might other factors be at work? Explain.

Chlorination of Toluene

Combustion of gasoline is assisted by "free radical initiators" such as tetraethyl lead, heating of which results in bond breaking and generation of ethyl radical.

$$(CH_3CH_2)_3Pb\text{-}CH_2CH_3 \xrightarrow{\Delta} (CH_3CH_2)_3Pb^{\bullet} + CH_3CH_2^{\bullet}$$

In the presence of chlorine gas and a free-radical initiator, three of toluene's hydrogens are sequentially replaced by chlorine atoms.

$$C_7H_8 \xrightarrow[\text{initiator}]{Cl_2} C_7H_7Cl \xrightarrow{Cl_2} C_7H_6Cl_2 \xrightarrow{Cl_2} C_7H_5Cl_3$$

The first step in the overall process is believed to involve abstraction of hydrogen by chlorine atom, followed by reaction of the ensuing radical with Cl_2.

$$C_7H_8 \xrightarrow{Cl^{\bullet}} C_7H_7^{\bullet} + HCl \xrightarrow{Cl_2} C_7H_7Cl + Cl^{\bullet}$$

Draw resonance structures for the possible radicals resulting from hydrogen atom abstraction from toluene. Which would you anticipate to be the most stable? Why? Compare energies for the different radicals (***radical A***, ***radical B***, ...). Is the lowest-energy radical that which you anticipated? Are any of the alternatives significantly better than any of the others? Explain your reasoning.

What are the three products resulting from free-radical chlorination of toluene? Why are only three hydrogens replaced?

Electron Flow in Diels-Alder Reactions

Electrostatic interactions play a significant role in determining the rates of Diels-Alder reactions.

Compare electrostatic potential maps for the following Diels-Alder transition states: ***cyclopentadiene+ethene***, ***cyclopentadiene+ acrylonitrile*** and ***cyclopentadiene+ tetracyanoethylene***, with those of reactants: ***cyclopentadiene***, ***ethene***, ***acrylonitrile*** and ***tetracyano-ethylene***. Recall that regions with excess negative charge will be colored red while those with excess positive charge will be colored blue. Are electrons transferred from diene to dienophile in the transition states (relative to reactants) or vice versa? For which reaction is the transfer the greatest? The least? Quantify your conclusion by measuring the total charge on the diene and dienophile components in the three transition states.

Calculate activation energies for the three Diels-Alder reactions (energy of transition state - sum of energies of reactants). Which reaction has the smallest energy barrier? Which has the largest energy barrier? Do your results parallel the measured relative rates of the same reactions (see table below)?

Experimental relative rates of Diels-Alder reactions involving cyclopentadiene

dienophile	relative rate
ethene	6×10^{-6}
acrylonitrile	1
tetracyanoethylene	4×10^{7}

Is there a correlation between activation energy and the magnitude of charge transfer between diene and dienophile components in the transition state? Explain.

Thermodynamic vs. Kinetic Control

Chemical reactions often yield entirely different product distributions depending on the conditions. High temperatures and long reaction times favor the most stable ("thermodynamic") products, while low temperatures and short reaction times favor the most easily formed ("kinetic") products. Consider Diels-Alder reaction of cyclopentadiene and maleic anhydride, leading to *endo* or *exo* adducts.

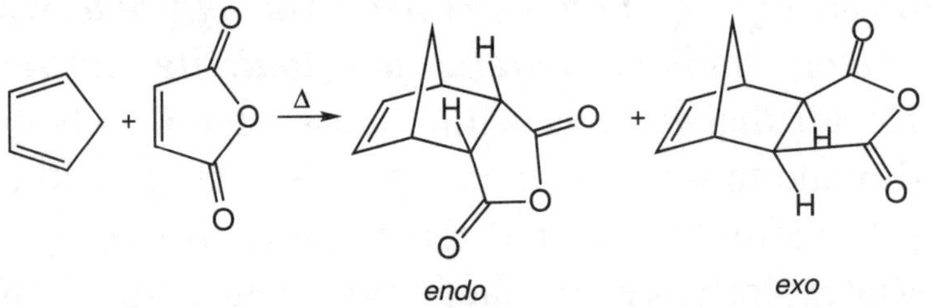

Display space-filling models of ***endo*** **and** ***exo adducts***. Which appears less crowded? What interactions disfavor the higher-energy adduct? Compare energies of the two adducts. Were the reaction under thermodynamic control, which would be the major product and what would be the ratio of products? Use the equation below.

$$\frac{N_{major}}{N_{minor}} = e^{-1060(E_{major}-E_{minor})}$$

N_i is the number of molecules *i*
E_i is the energy of molecule *i* (in au)

Compare energies of ***endo*** **and** ***exo transition states***. Were the reaction under kinetic control, which would be the major product and what would be the ratio of products? Use the equation below.

$$\frac{N_{major}}{N_{minor}} = e^{-1060(E^{\ddagger}_{major}-E^{\ddagger}_{minor})}$$

N_i is the number of molecules *i*
E_i is the energy of the transition state leading to molecule *i* (in au)

Are the kinetic and thermodynamic products the same? If not, describe conditions which will favor the *endo* adduct. The *exo* adduct.

Addition vs. Substitution

Unsaturated hydrocarbons undergo a variety of reactions. Experimentally, alkenes and alkynes undergo addition reactions, whereas aromatic molecules, such as benzene, undergo substitution reactions instead. Why?

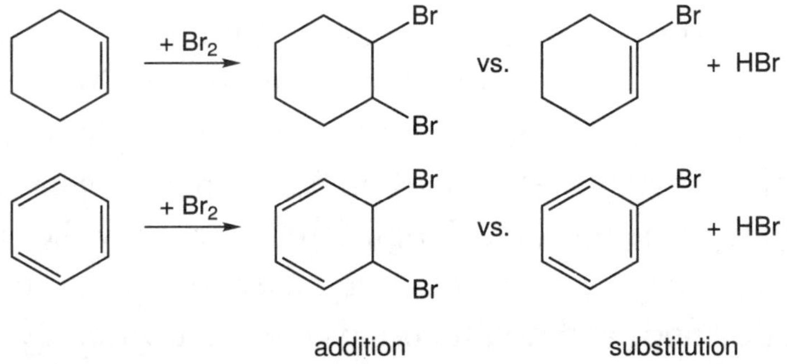

Calculate the energy of addition of bromine (E=-5120.5302 au) given at left) to ***cyclohexene*** leading to ***trans-1,2-dibromocyclohexane***. Is this reaction exothermic? Next, calculate the energy of the corresponding substitution reaction, leading to ***1-bromocyclohexene*** and hydrogen bromide (E=-2560.8428 au). Is this reaction exothermic? Do the thermodynamics of these alternative reactions account for which pathway is followed? Explain.

Repeat this analysis for the bromine addition and substitution reactions of ***benzene*** leading to ***trans-5,6-dibromo-1,3-cyclohexadiene*** and ***bromobenzene***, respectively. Do your thermochemical results account for the experimental observations?

What aspects, if any, of cyclohexene and benzene reaction thermodynamics are similar? Why do you suppose this is? What aspects are different? Why?

Hückel's Rule. Cyclooctatetraene

Hückel's rule states that planar cyclic π systems involving 4n+2 electrons will be unusually stable ("aromatic"), while cyclic π systems with 4n electrons will be unstable ("antiaromatic"). This suggests that cyclooctatetraene (C_8H_8) should be quite unlike benzene (C_6H_6).

Compare energies of ***planar*** and ***tub cyclooctatetraene***. Which is lower? Is the higher-energy form an energy minimum? To tell, examine its vibrational frequencies. Recall that energy minima have all real frequencies, while molecules with one or more imaginary frequencies are not minima. Are all carbon-carbon bonds in the lower-energy form of cyclooctatetraene the same length? If so, are they the same length as those in ***benzene***? If not, do they alternate between "normal" carbon-carbon single bonds and double bonds (*see also Page 29*)?

Display the electrostatic potential map for the lower-energy form of cyclooctatetraene. Recall that the most electron-rich regions will be colored red. Where is the highest concentration of negative charge?

Is cyclooctatetraene aromatic? To tell, compare the first and second hydrogenation energies, leading to ***1,3,5-cyclooctatriene*** and then to ***1,3-cyclooctadiene***. (The energy for hydrogen is 0 au.)

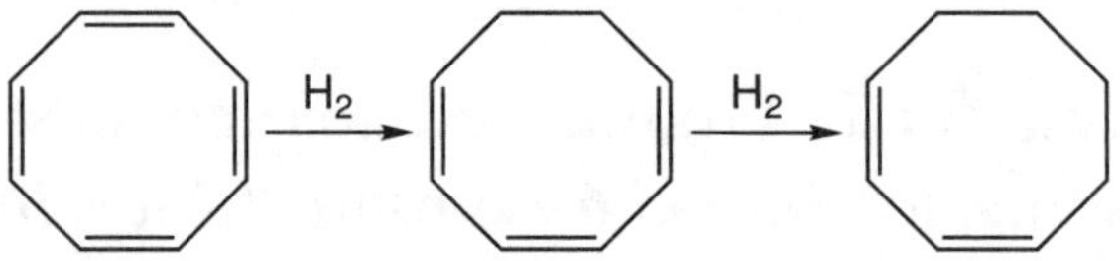

Whereas the initial hydrogenation both breaks a π bond and destroys any "aromatic stabilization," the second hydrogenation only breaks a π bond. The difference between the two then corresponds to any aromatic stabilization. For comparison, hydrogenation of benzene to 1,3-cyclohexadiene is endothermic by .001 au (6 kcal/mol), while hydrogenation of 1,3-cyclohexadiene to cyclohexene is exothermic by .041 au (26 kcal/mol). The difference .051 au (32 kcal/mol) is one measure of the aromaticity of benzene.Is this difference large as in benzene or is it neglible? Is cyclooctatetraene aromatic?

Nucleophilicity of Benzene and Pyridine

Benzene and substituted benzenes react with electrophiles, leading to new functionality. The two-step mechanism involves initial attack by an electrophile to form an intermediate (benzenium ion), followed by elimination of a "proton" to generate the substituted benzene.

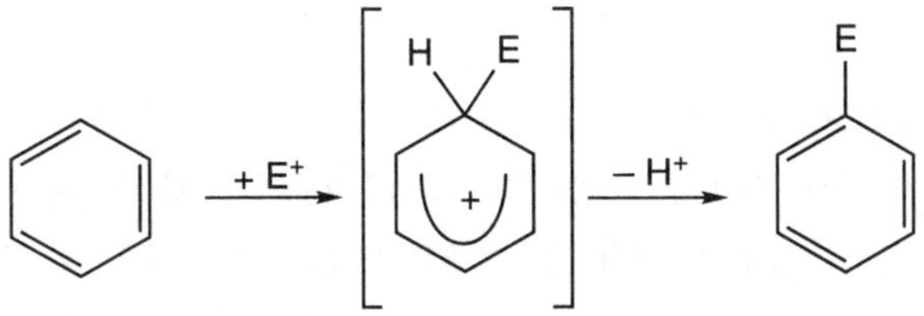

benzenium ion

Display the electrostatic potential map for ***benzene***. Which areas are most electron rich (most red)? Which are most electron poor (most blue)? Would you expect an electrophile to attack from above and below the plane of the molecule or in the plane of the molecule?

Draw and compare Lewis structures for benzene and pyridine. How many π electrons does each molecule have? Where are the most accessible electrons in each? Display the electrostatic potential map for ***pyridine*** and compare it to the corresponding map for benzene. Would you expect electrophilic attack on pyridine to occur analogously to that in benzene? If so, should pyridine be more or less susceptible to aromatic substitution than benzene? If not, where would you expect electrophilic attack to occur? Explain.

Imidazole and Pyrazole. Where is the Basic Site?

Both imidazole and pyrazole are moderately strong bases.

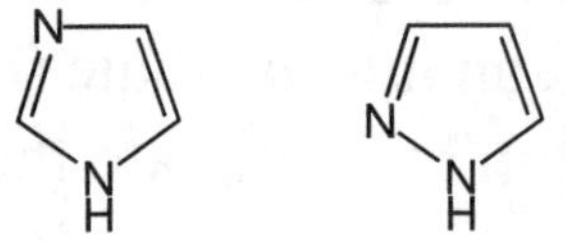

imidazole pyrazole

Draw a Lewis structure for each molecule that shows the location of all nonbonding electrons. Examine electrostatic potential maps for both ***imidazole*** and ***pyrazole***. Recall that regions of excess negative charge (most likely protonation sites) are colored red. Predict which is the more basic nitrogen in each molecule. What kind of orbital contains this nitrogen's nonbonding electrons? What kind of orbital contains the other nitrogen's nonbonding electrons?

Which nitrogen in each molecule is more basic? Compare energies of the alternative conjugate acids (***N protonated imidazole***, ***NH protonated imidazole***, ***N protonated pyrazole*** and ***NH protonated pyrazole***). Which compound, imidazole or pyrazole, is more basic? Compare the energies of protonation (leading to the favored conjugate acid in each case). Rationalize your result.

Polar Hydrocarbons

Neutral hydrocarbons are generally nonpolar molecules. This is to be expected since carbon-carbon and carbon-hydrogen bonds are relatively nonpolar. Resonance effects can alter this picture, however, by redistributing electrons in novel ways.

Examine electrostatic potential maps for ***naphthalene***, ***azulene*** and ***hexaphenyltriafulvene***. Recall, that regions of excess negative charge will be colored red while regions of excess positive charge will be colored blue.

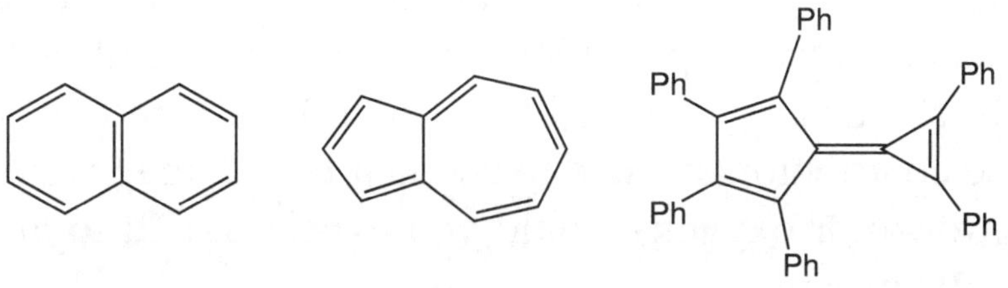

naphthalene azulene hexaphenyltriafulvene

Identify the most negatively-charged and most positively-charged regions in each molecule. (Ignore the phenyl rings attached to triafulvene.) The dipole moments of these molecules have been measured as 6.3, 0 and 0.8 debyes. Which molecule is responsible for which dipole moment? Explain the trend in dipole moments.

Each molecule above contains two conjugated rings. According to Hückel's rule, how many π electrons must each ring have in order for it to be aromatic? Which molecules must transfer π electrons from one ring to the other in order to become aromatic? Draw resonance contributors that show this electron transfer.

Some triafulvene derivatives undergo rotation about the carbon-carbon double bond even at room temperature. Given that *cis-trans* isomerization is normally very difficult, how would you rationalize this? Examine the electrostatic potential map for ***perpendicular hexaphenyltriafulvene*** (the rotational transition state). Would polar solvents tend to lower or raise the rotation barrier? Explain.

Useful Electrophiles

Among the most common and synthetically-useful electrophiles are nitronium and acyl cations, NO_2^+ and CH_3CO^+, respectively. The former is the active agent in electrophilic nitration while the latter is the active reagent in Friedel-Crafts acylation.

Examine the geometry of ***nitronium cation***. Is it linear or bent? Draw the ion's Lewis structure. What common neutral organic molecule is isoelectronic with NO_2^+? Is this molecule linear or bent? Examine the charges on the nitrogen and oxygen atoms in NO_2^+. Is nitrogen or oxygen more positive? Does this agree with your Lewis structure? Display an electrostatic potential map for NO_2^+. Recall that the most positively-charged regions will be colored blue. Are any additional resonance contributors needed to account for the map results? What atom (nitrogen or oxygen) would you expect to add to an arene in electrophilic nitration?

Examine charges and the electrostatic potential map for ***acyl cation***. Which atom is the most positively charged? Draw the Lewis structure for acyl cation that is most consistent with its charge distribution. Is the calculated geometry of acyl cation consistent with its Lewis structure? Hint: Compare CC and CO bond distances to "typical" single, double and triple values given below.

CC and CO bond distances (Å)

C–C	1.54
C=C	1.32
C≡C	1.20
C–O	1.35
C=O	1.22
C≡O	1.12

What common neutral organic molecule is isoelectronic with acyl cation? Is the structure of this molecule similar to CH_3CO^+? Which atom (carbon or oxygen) would you expect to add to an arene in electrophilic acylation?

Directing Effects on Electrophilic Nitration

Electrophilic nitration of a substituted benzene may lead to *ortho*, *meta* or *para* products, depending on the substituent. According to the Hammond Postulate, the kinetic product will be that which follows from the most stable intermediate benzenium ion, i.e.

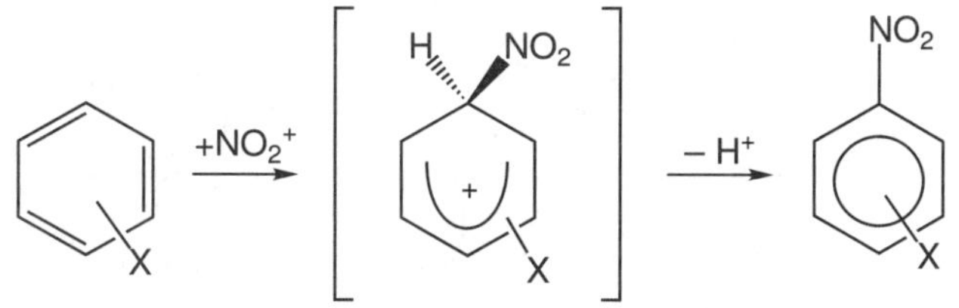

Draw a Lewis structure (or a series of Lewis structures) for nitrobenzenium ion. Where is the positive charge? Examine the electrostatic potential map for ***nitrobenzenium ion***. Recall that the most positively-charged regions will be colored blue. Where would you expect electron-donor substituents to have the greatest stabilizing effect (consider *meta* and *para* positions only)? Which is the more stable, ***meta*** or ***para-nitrotoluenium ion*** (intermediates in nitration of toluene)? Compare electrostatic potential maps to that for nitrobenzenium ion. Does your result suggest that methyl acts as an electron donor?

Compare energies for ***meta*** and ***para-nitroanilinium ions*** (intermediates in nitration of aniline). Are these differentiated to a lesser or greater extent than the intermediates in toluene nitration? Examine electrostatic potential maps. What do these suggest about the relative electron-donor strengths of methyl and amino groups?

Compare energies for ***meta*** and ***para-dinitrobenzenium ions*** (intermediates in nitration of nitrobenzene). Is the ordering the same as those observed for intermediates in toluene and aniline nitration? Examine electrostatic potential maps. What does your result suggest about the electron donor/acceptor properties of the nitro substituent?

Predict the nitration products of toluene, aniline and nitrobenzene.

Wade Section 17-12A

Nucleophilic Aromatic Substitution. Addition-Elimination

Aryl halides undergo substitution, although not through an S_N2 mechanism, but rather via a two-step "addition-elimination" mechanism. (An "elimination-addition" mechanism is also possible.)

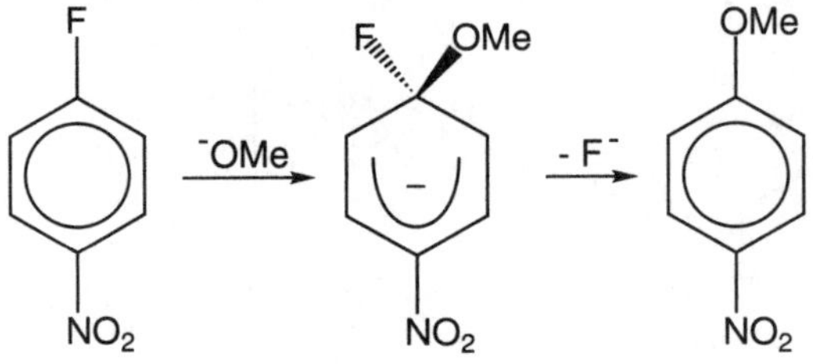

Display electrostatic potential maps for ***fluorobenzene***, ***para-nitrofluorobenzene*** and ***2,4,6-trinitrofluorobenzene***. Recall that the blue color designates regions of highest positive charge and most subject to nucleophilic attack. Which should be the most susceptible toward nucleophilic attack? Which should be the least susceptible? Explain.

Draw Lewis structures for the anion resulting from addition of methoxide to fluorobenzene. Is the negative charge highly localized or is it delocalized over several positions? Examine calculated charges for the anion. Are they anticipated from the Lewis structures?

Is *para*-nitrofluorobenzene more or less susceptible to attack by methoxide than fluorobenzene? Calculate the energetics of the reaction. (Energies for ***fluorobenzene*** and ***para-nitrofluorobenzene methoxide anion adducts*** are available.)

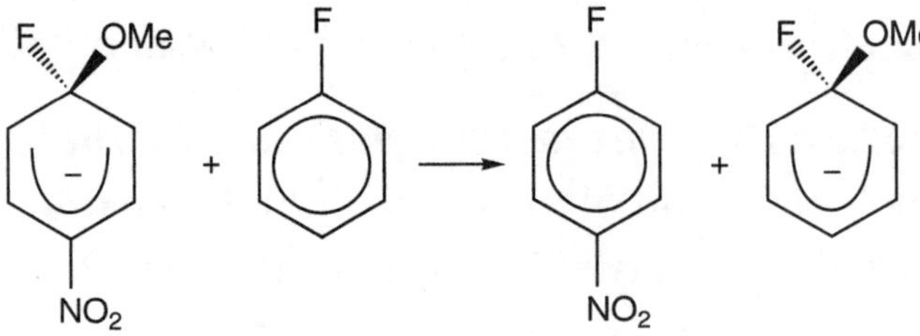

Does your result suggest that nitro is acting as an electron-donating or electron-withdrawing group? Explain.

Nucleophilic Aromatic Substitution. Benzyne

In a commercially important synthesis, aqueous sodium hydroxide reacts with chlorobenzene to give phenol.

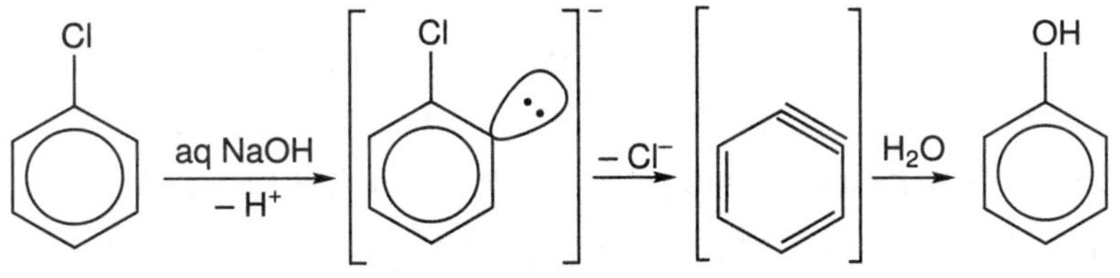

The "addition-elimination" mechanism involves two intermediates, a chlorophenyl anion and benzyne. A simple displacement mechanism can be ruled out because reaction of *ortho*-chlorotoluene gives not only *ortho*-methylphenol but also *meta*-methylphenol.

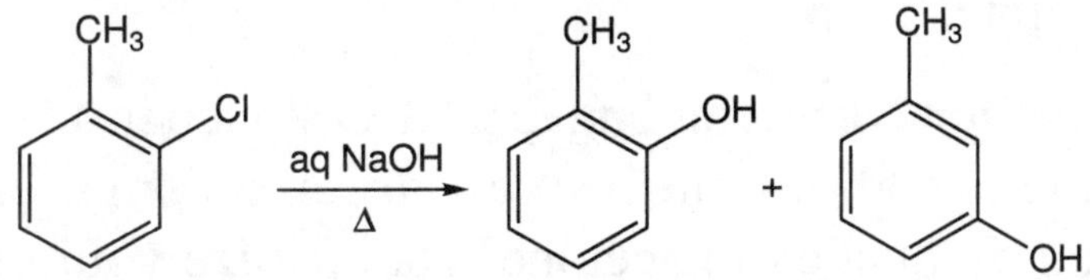

Examine the geometry of ***methylbenzyne***. Measure carbon-carbon distances. Which π bonds are delocalized and which are localized? Is there really a triple bond? (Compare bond distance to triple bond in ***hexa-1,5-dien-3-yne*** and to partial double bonds in ***benzene***). Are you able to draw a single Lewis structure which adequately represents the geometry of the molecule?

Hydration of methylbenzyne is believed to require nucleophilic attack by hydroxide. Examine the lowest-unoccupied molecular orbital (LUMO) of methylbenzyne. This should reveal likely sites for nucleophilic attack. How many sites are there for nucleophilic attack? Does hydroxide attack in the plane of the ring, or perpendicular to the ring plane? Explain.

Infrared Spectra of Carbonyl Compounds

The CO stretch in the infrared spectra of carbonyl compounds gives rise to a strong absorption around 1700 cm^{-1}, and is often used as a diagnostic for this functional group.

Examine each of the vibrational motions for ***acetone***, and identify the motion corresponding to the CO stretch. Note, that calculated vibrational frequencies are typically about 12% larger than measured frequencies. For example, a measured frequency of 1700 cm^{-1} would correspond to a calculated frequency of around 1900 cm^{-1}. What is the CO stretch frequency? Are there any other vibrations which have very similar frequencies? What does this suggest about the use of the CO stretching frequency as a diagnostic for carbonyl compounds? Elaborate. Does the "CO stretching frequency" involve significant motion of any atoms other than the two which make up the carbonyl group? Elaborate.

Identify the motion corresponding to the CO stretch in ***acetophenone***. Is the frequency about the same, larger or smaller than the corresponding frequency in acetone? Rationalize your observation. Hint: Draw contributing resonance structures for acetophenone. Would you expect the CO bond to be about the same, longer or shorter than that in acetone? Compare bond lengths in acetone and acetophenone to check your reasoning. Is there a relationship between CO bond length and CO stretching frequency? If so, try to rationalize.

Compare the geometry of ***2,6-dimethylacetophenone*** to that of acetophenone. In particular, consider the orientation of the phenyl ring and the carbonyl group. Is one of the molecules better suited than the other for conjugation? Next, identify the CO stretching frequency in 2,6-dimethylacetophenone. Is it significantly smaller or significantly larger than that in acetophenone, or is it about the same? What, if anything, does this tell you about the importance of conjugation in these systems?

Push-Pull Resonance. The Basicity of *para*-Nitroaniline

para-Nitroaniline is a much weaker base than aniline, an observation which is usually explained by invoking so-called "push-pull" resonance contributors.

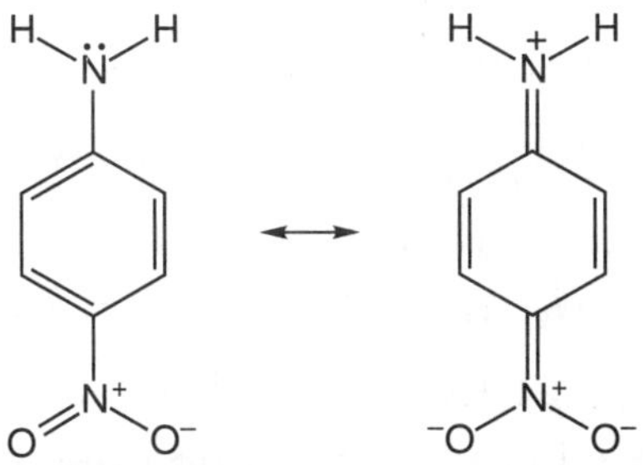

Compare the geometry of ***para-nitroaniline*** to those of both ***aniline*** and ***nitrobenzene***. Is there any evidence for push-pull resonance contributors? Is there shortening of bonds to the amino and nitro groups? Are the bonds in the ring localized? Is the dipole moment for *para*-nitroaniline smaller, larger or about the same as the sum of the dipole moments for aniline and nitrobenzene? What does your result say about the importance of push-pull resonance contributors?

Compare electrostatic potential maps for *para*-nitroaniline and aniline. Recall that regions of excess negative charge will be portrayed in red and regions of excess positive charge portrayed in blue. Has the NO_2 group decreased, increased or left unchanged the electrostatic potential at the amino group? What effect, if any, should this have on basicity? Also examine electrostatic potential maps for ***para-methylaniline*** and ***para-trifluoromethylaniline***. Would you predict these to be stronger or weaker bases than aniline? Rationalize your results using changes in geometries and/or changes in dipole moments. Data for ***toluene*** and ***trifluorotoluene*** are available.

Phase-Transfer Catalysis

Phase-transfer catalysis describes the action of special catalysts that assist the transfer of reactive molecules from a polar ("aqueous") solvent to a nonpolar ("organic") solvent. In the absence of the phase-transfer catalyst, one of the reagents is confined to one solvent, and the other reagent is confined to the other solvent, so no reaction occurs. Addition of a small amount of catalyst, however, enables one of the reagents to pass into the other solvents thereby initiating a reaction.

For example, cyclohexene, chloroform and aqueous NaOH react only when combined with a phase-transfer catalyst, such as benzyltriethylammonium chloride. The organic reagents stay in the organic layer (usually chloroform), but OH^- is able to migrate between the aqueous and organic layers if accompanied by the catalyst (NaOH is insoluble in chloroform). The resulting reaction involves deprotonation of chloroform by OH^- giving dichlorocarbene, CCl_2, which then combines with the alkene.

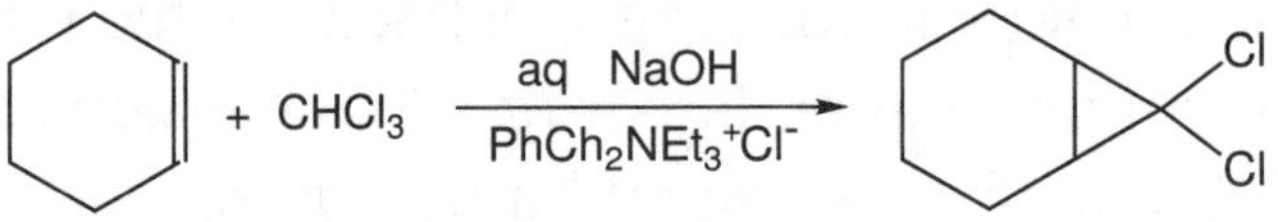

What properties of ***benzyltriethylammonium ion*** make it soluble in diverse solvents? Examine its electrostatic potential map and atomic charges. Which groups facilitate water solubility? Explain. Which groups facilitate chloroform solubility? Explain.

Compare electrostatic potential maps for ***tetrabenzylammonium ion*** and ***tetraethylammonium ion*** with that of benzyltrimethylammonium ion. Recall that the most positively-charged regions will be colored blue. Are they likely to be as effective or more effective as phase-transfer catalysts as benzyltrimethylammonium ion? Explain. (Hint: Predict solubility properties for the three ions.)

Intra and Intermolecular Hydrogen Bonding

The melting and boiling points of carboxylic acids are much higher than would be expected on the basis of their molecular weights. The usual explanation is that they form weak intermolecular bonds.

Display an electrostatic potential map for ***acetic acid***. Recall that electron-rich sites which might serve as hydrogen-bond acceptors will be colored red, while electron-poor sites which might serve as hydrogen bond donors are colored blue. Where are the most electron-rich sites? Where are the most electron-poor sites? Propose a structure for the dimer of acetic acid based on favorable electrostatic interactions between electron-rich and electron-poor sites. Compare your structure to that for ***acetic acid dimer***. What is another name for the types of interactions that hold the two acetic acid molecules together?

The melting points of open-chain dicarboxylic acids, $H_2OC(CH_2)_nCO_2H$, are also typically higher than might be expected on the basis of molecular weight alone. Examine the structure of ***nonane-1,9-dioic acid***. (Molecules of the complexity open-chain dicarboxylic acids typically exist as a collection of many different conformers, and the conformer displayed corresponds to the calculated lowest-energy structure.) How is it similar and how is it different from the structure of acetic acid dimer? Account for the high melting point of the compound.

Vitamin C. Ascorbic Acid

Organic acids usually contain a carboxylic acid group, -CO_2H. L-Ascorbic acid, commonly known as Vitamin C, is an obvious exception.

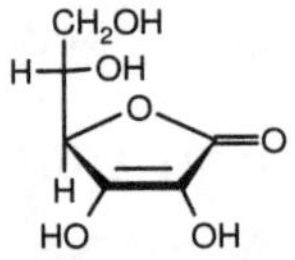

Examine atomic charges as well as the electrostatic potential map of ***ascorbic acid***. Which hydrogen(s) is likely to be most acidic? This is an electron-poor site and should be colored blue.

Obtain the energies of the various ***conjugate bases*** of ascorbic acid. Which one is the most stable? Is it the base which results from deprotonation of the hydrogen you previously assigned as the most acidic?

Examine the structures and atomic charges for the various conjugate bases. How do they differ? What distinctive features, if any, characterize the most stable conjugate base? Draw all of the resonance contributors needed to account for the electron distribution and geometry of the most stable conjugate base.

Enolate Acidity, Stability and Geometry

In general, alkyl hydrogens are not very acidic. However, alkyl hydrogens adjacent to carbonyl groups can be deprotonated by strong bases to give enolate anions, e.g., for acetone.

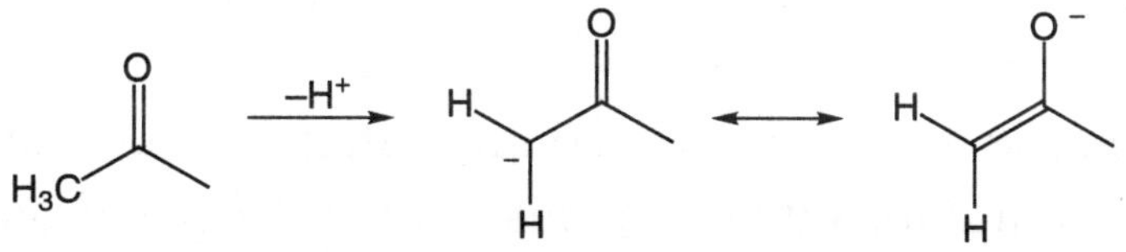

Compare electrostatic potential maps for ***propane***, ***acetone*** and ***2,4-pentanedione***, and identify the most positively-charged "acidic" hydrogen(s) in each. These will be colored blue. Is there a correlation between electrostatic potential and pK_a (see table below)?

	pK_a
propane	50
acetone	20
2,4-pentanedione	9

How many different enolates may arise from deprotonation of 2,4-pentanedione? Draw Lewis structures for each, and predict which is likely to be the most stable. Check your conclusions by examining the energies of the different possible enolates (***enolate A***, ***B***...). Is the most stable enolate that derived from deprotonation of the most electron-poor hydrogen? Compare the electrostatic potential maps of the anions with each other and with your Lewis structures. The most negatively-charged regions will be colored red. Revise your drawings to be consistent with the maps. Why is one of the enolates preferred over the others?

Wade Section 22-7

Aldol Condensation

Aldehydes and ketones are deprotonated by strong bases to give enolates. These can then react with a second molecule and eliminate water, for example, reaction of acetaldehyde.

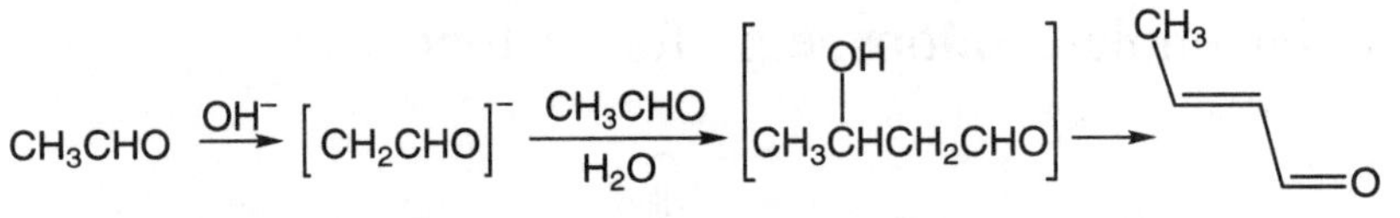

For unsymmetrical ketones, two condensation products are possible. For example, intramolecular condensation of 2,7-octadione may lead to products which follow from the two possible enolates.

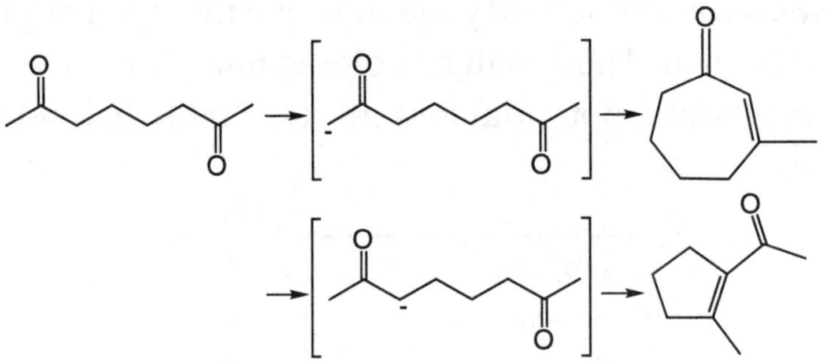

Examine atomic charges and display the electrostatic potential map for ***2,7-octadione***. Recall that the most positively-charged regions (the likely acidic sites) will be colored blue. Are you able to say which hydrogens (at C_1 or at C_3) are more likely to be abstracted by base, and conclude which is the kinetically-favored enolate? Which enolate (***2,7-octadione***, ***C1 enolate*** or ***C3 enolate***) is the lower in energy? What do you conclude is the thermodynamically-favored enolate? Is this also the enolate in which the negative charge is better delocalized? Compare electrostatic potential maps to tell.

Finally, examine the transition states for ***closure*** of the C_1 enolate ***to*** the ***7-membered ring*** product, and of the C_3 enolate ***to*** the ***5-membered ring*** product. Calculate activation energy barriers from their respective enolates. Which ring closure (to the five or the seven-membered ring) occurs more readily? Overall, what do you conclude is the dominant condensation product of 2,7-octadione?

Michael Addition

α,β-Unsaturated carbonyl compounds may undergo nucleophilic addition either at the carbonyl carbon (carbonyl addition) or at the β carbon (Michael addition), thus leading to different products, e.g., in cyclohexenone.

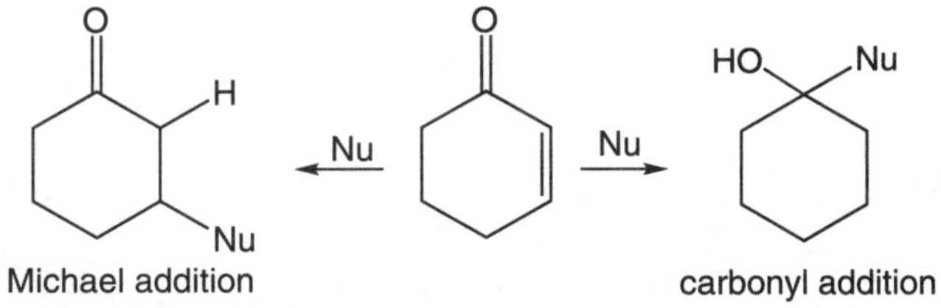

Draw a Lewis structure for cyclohexenone that involves charge separation for the most polar bond. Then, draw a Lewis structure that will delocalize one or both charges. Next, examine the actual geometry of ***cyclohexenone***. Are the bond distances consistent with the Lewis structure shown above, or have they altered in accord with your alternative (charge separated) Lewis structure? (Structures for ***cyclohexene*** and ***cyclohexanone*** are available for reference.)

Display and describe the lowest-unoccupied molecular orbital (LUMO) for cyclohexenone. This is the orbital into which the nucleophile's pair of electrons will go, that is, the likely site of nucleophilic attack. Does it anticipate both carbonyl and Michael products of nucleophilic addition? Explain.

Glucose

Glucose undergoes a large number of selective, acid-catalyzed substitution reactions at C_1. Since the role of acid is to protonate the C_1 hydroxyl and make it a better leaving group, it is not apparent why only this hydroxyl group gets replaced.

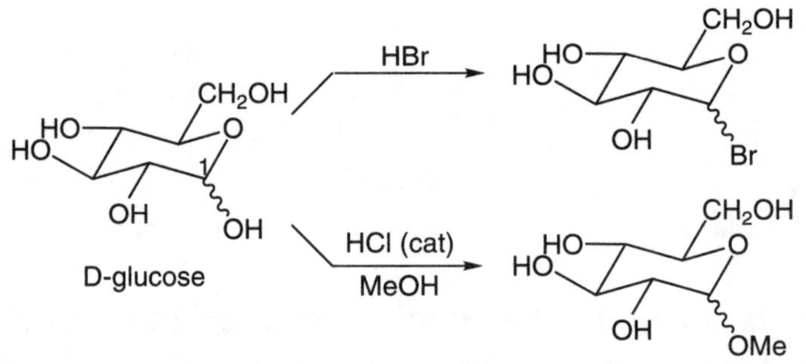

Examine the electrostatic potential map of ***beta-D-glucose*** and identify all of the electron-rich sites. These are colored red and should correspond to the likely protonation sites. Does one site appear to be significantly more electron rich than the others, or do several sites appear to have comparable reactivity?

Protonation and subsequent loss of water should generate a carbocation. Examine all of the ***carbocations*** derived from protonation of β-D-glucose. Identify the most stable carbocation (this is the one that will form most readily), and draw whatever resonance contributors are needed to describe the geometry, energy, and atomic charges in this cation. Can you explain why substitution occurs selectively at C_1?

Lactose undergoes acid hydrolysis to give glucose and galactose. If the bridging O (*O) were labeled with ^{18}O, would you expect this label to be found in glucose or in galactose?

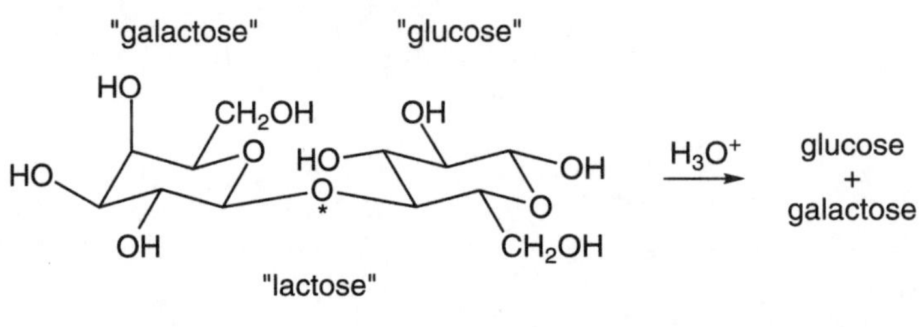

DNA Base Pairs

The two strands which make up DNA are held together by hydrogen bonds between complementary pairs of bases: adenine paired with thymine and guanine paired with cytosine. The integrity of the genetic code (and of life as we know it) depends on error-free transmission of base-pairing information.

Examine electrostatic potential maps for ***adenine***, ***thymine***, ***guanine*** and ***cytosine*** (A, T, G and C, respectively). Electron-rich region, which may act as hydrogen bond acceptors, are colored blue, while electron-poor regions, which may act as hydrogen-bond donors, are colored blue. Identify potential hydrogen-bond donor and acceptor sites for each. Sketch possible hydrogen-bonded geometries for the adenine-thymine pair and for the guanine-cytosine pair. How many hydrogen bonds hold together adenine and thymine? How many hold together guanine and cytosine?

Examine ***AT pair*** and ***GC pair***, adenine-thymine and guanine-cytosine base pairs, respectively. Identify the hydrogen bonds in each. Are they the same as those you sketched? Are the base pairs "flat" as normally drawn in textbooks, or are they significantly "puckered" or "twisted"?

Calculate binding energies for both the adenine-thymine and guanine-cytosine base pairs (energy of the base pair less the energies of the component bases). Also calculate binding energies for "incorrect" base pairs (***AG pair*** and ***TC pair***). Estimate the "energetic incentive" for base pairing to occur "correctly."

Structure of the Double Helix

DNA is made up of two intertwined strands. A sugar-phosphate chain makes up the backbone of each, and the two strands are joined by way of hydrogen bonds betwen pairs of nucleotide bases, adenine, thymine, guanine and cytosine. Adenine may only pair with thymine and guanine with cytosine (*see Page 83*). The molecule adopts a helical structure (actually, a double helical structure or "double helix").

Examine ***DNA***. How many base pairs does it contain? Starting from one end, write down the sequence of bases in one strand. Write down the sequence in the complementary strand. Is this a "proper" DNA fragment, or does it contain base-pair mismatches?

How many base pairs are required for a full (360°) turn of the helix? Locate the minor and major grooves in the DNA fragment. Will a polycyclic aromatic hydrocarbon such as ***anthracene*** be able to fit into the major groove? Examine space-filling models to tell.

Structure of Glycine in the Gas Phase and in Water

Glycine and other amino acids are usually thought of as zwitterions, bearing a formal positive charge at nitrogen and a formal negative charge at oxygen.

$$H_3N^+—CH_2—CO_2^-$$

An alternative involves uncharged amino and carboxylic acid groups.

$$H_2N—CH_2—CO_2H$$

"Zwitterion" and "non-zwitterion" isomers are related by the shift of a proton, and are known as tautomers.

To what extent do zwitterions bear "full" + and - charges? Compare atomic charges in the ***zwitterionic*** and ***non-zwitterionic*** forms of ***glycine***. What is the total charge on -NH_3 in the zwitterionic form? What is the total charge on -CO_2? Are these charges much greater than those on -NH_2 and -CO_2H, respectively, in the non-zwitterionic form of glycine? Is the dipole moment for zwitterionic glycine much greater than that for the non-zwitterionic tautomer? Display electrostatic potential maps for both zwitterionic and non-zwitterionic glycine. Recall that regions of excess negative charge will be colored red, while regions of excess positive charge will be colored blue. Which shows the greater charge separation?

Which form of glycine, zwitterionic or non-zwitterionic, is the lower-energy species in the gas phase? Rationalize your observation. Does the ordering or the difference in tautomer stabilities change in going to an aqueous environment? Structures, atomic charges and electrostatic potential maps for both ***zwitterionic*** and ***non-zwitterionic*** forms of ***glycine*** surrounded by 20 ***water*** molecules are available. Which is the lower energy form? Has solvation had a greater effect on atomic charges and electrostatic potentials for the zwitterionic or non-zwitterionic form? Account for your observation.

Amino Acid Sidechains

The 20 natural amino acids differ from each other by the nature of their sidechains. Differences involve overall size, hydrophobic or hydrophilic character and, perhaps most importantly, ionization state. The latter is characterized by an "isoelectric point," the pH at which it exists in neutral form. Differences in isoelectric points may be exploited to separate amino acids in what is termed an electrophoresis experiment. While the sidechains are normally written in terms of "neutral" structures, some may also exist in either protonated or deprotonated forms depending on pH.

A selection of amino acids (***acid A***, ***acid B***,...) terminated at both ends by amide functionality, i.e., MeNHCO–CHR–NHCOMe, are provided. These are given in the ionization states found at neutral pH. For each, first identify the amino acid, and then the ionization state (neutral, protonated or deprotonated). Next compare electrostatic potential maps among the different amino acids. Recall that negatively and positively charged regions, which would favor hydrophilic environments, will be colored red and blue, respectively, while "neutral" (hydrophobic) regions will be colored green. Which amino acids would prefer hydrophobic environments? Hydrophilic environments? Explain your reasoning.

The X-ray crystal structures of proteins show that highly polar and/or charged amino acids usually congregate on the exterior regions while less polar, uncharged amino acids congregate in interior regions. Explain. For each of the amino acids above, indicate a preference for interior or exterior regions.

Structure of Polypeptides

Polypeptides (or more simply peptides) are amino acid "polymers" comprising up to approximately 50 units.

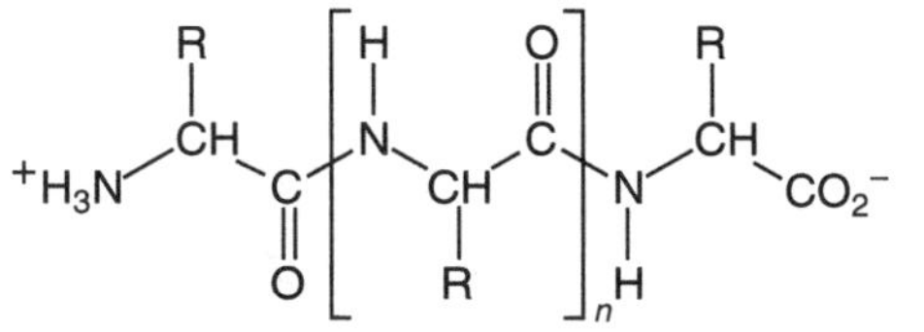

Shown above is the so-called zwitterionic form with charged NH_3^+ and CO_2^- terminal groups. This is the favored structure in aqueous solution. A "neutral" form, with NH_2 and CO_2H terminal groups, is favored in the gas phase (see also, **Problem 24-3**). Larger "polymers" are known as proteins. Aside from the amide linkages, the polymer chain is very flexible, giving rise to the possibility of an enormous number of different conformers. It is nothing short of remarkable that proteins rapidly fold into a single conformation. Very strong forces must be at work.

Two of the most common motifs in the hundreds of known protein structures are the so-called α helix and the β sheet. Examine ***alpha helix***, a short "helical" strand of glycine units. How many glycines make up the strand? How many glycines are required for a full turn of the helix? Does incorporation into a helix cause the amide groups to distort significantly away from their "ideal" planar arrangements? What kind of "bonds" hold the helix together? How many, if any, of these bonds can you identify in a full turn? Look down on the helical axis. Is there a cavity? Display the molecule as a space-filling model. Is there still a cavity?

Examine ***beta sheet***, a structure made up of two parallel strands of glycine units. How many glycines are in each strand? What kind of "bonds" hold this structure together? Is the number of these bonds (per glycine) less, greater or the same as for the α helix? Display the molecule as a space-filling model. Describe its shape. Are there any empty spaces? What effect might replacing glycine with another amino acid have on the stability of this structure? (Other amino acids have CHR whereas glycine has CH_2.)

Spin Traps and Radical Scavengers

The hydroxyl hydrogen in phenol is particularly susceptible to abstraction by a free radical.

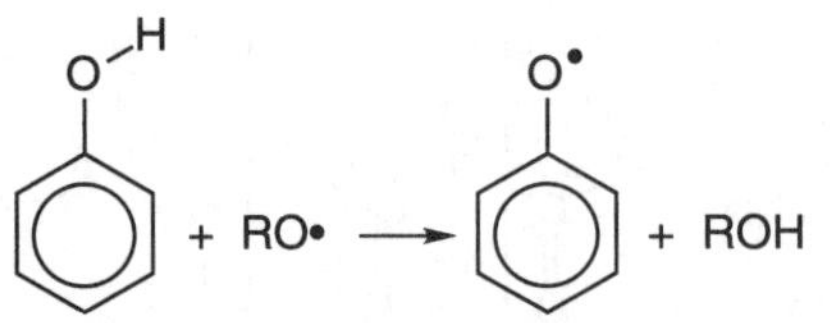

The process is exothermic, suggesting that the phenoxy radical is particularly stable. Display the spin density surface for ***phenoxy radical.*** This shows the location of the unpaired electron. Is it localized or delocalized over several centers? Is the unpaired electron in the σ or π system? Draw appropriate Lewis structures that account for your data.

Phenol is a "radical scavenger." Other radical scavengers include 3,5-di-*tert*-butyl-4-hydroxytoluene (BHT) and vitamin E.

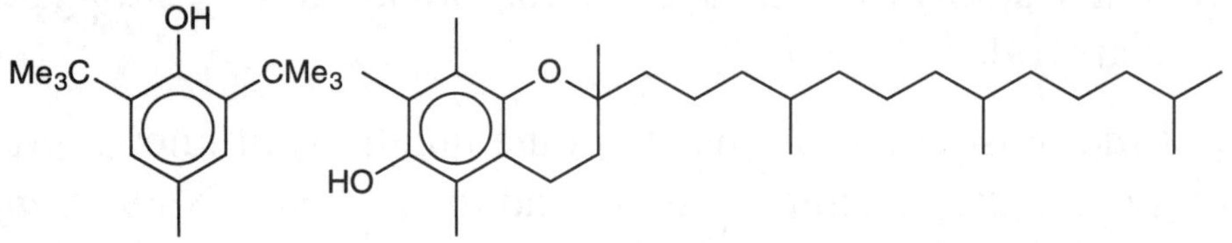

Examine the spin density surface for ***BHT radical.*** Is the unpaired electron localized or delocalized? Use a space-filling model to explain why BHT radical does not readily add to alkenes or abstract hydrogens from other molecules.) Compare the spin density surface for ***vitamin E radical*** to those of phenoxy and BHT radicals. Elaborate any significant differences among the three. What is the function of the long alkyl chain in vitamin E? Examine an electrostatic potential map for vitamin E radical. Recall that charged regions (colored red and blue for negative and positive regions, respectively) are likely to be compatible with polar environments, while uncharged or neutral regions are more likely to be compatible with non-polar environments. Do you expect vitamin E to be soluble in aqueous (polar) or non-aqueous (non-polar) environments, or both?

Fatty Acids and Fats. What Makes Good Soap?

Natural fats are glycerol esters of fatty acids known as triglycerides. Unsaturated fats are generally liquids (oils) at room temperature, while triglycerides rich in saturated fatty acids are generally solids. Examine models of ***tristearin*** and ***triolein***. You should recognize that molecules of this complexity typically exist as a collection of many different conformers, and those shown should only be taken as representative (low-energy) conformers. Which one of these is saturated and which is unsaturated? Are the double bonds in the unsaturated fat *cis* or *trans*?

Hydrolysis of animal fats in the presence of strong base leads to glycerol and salts of long-chain carboxylic acids. The latter are known as "soaps."

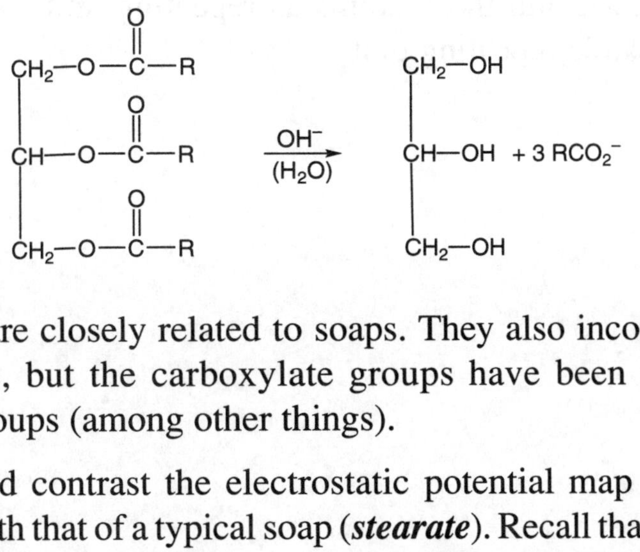

Detergents are closely related to soaps. They also incorporate long alkyl chains, but the carboxylate groups have been replaced by sulfonate groups (among other things).

Compare and contrast the electrostatic potential map of a typical ***detergent*** with that of a typical soap (***stearate***). Recall that negatively-charged regions are colored red, while "neutral" regions are colored green. Which part of each molecule will be most water soluble (hydrophilic)? Draw a Lewis structure that describes each molecule's water-soluble group (make sure you indicate all necessary formal charges and lone pairs). Which part(s) of each molecule will be most grease soluble (lipophilic)? What kinds of atoms and bonds are found in these groups?

Synthetic Polymers

A wealth of important materials fall under the general category of synthetic polymers. All share a common theme of being made up of sequences of one or more monomer units.

One after the other, examine the structures of a number of common ***monomers***. What features, if any, do they have in common? What relevance is this to the polymerization process?

One after the other, examine the structures of a number of common ***polymers***. No single conformer of a polymer will adequately represent its overall size and shape. The low-energy conformer for each polymer strand depicted is merely meant to allow identification of the polymer in terms of its components. For each, draw the repeating unit, and indicate the chain length (number of repeating units in the strand). Note: Each end of a polymer strand has been capped by adding extra atoms. Do not count these atoms as repeating units. Also, use the smallest possible repeating unit.

V. SPARTANBuild. An Electronic Modeling Kit

SPARTANBuild for Power Mac's and PC's (Windows 95/98/NT) allows you to construct a wide variety of organic molecules, including the major types of reactive intermediates. It serves the same purpose as the set of "plastic" models commonly used among organic chemistry students, although it is far more powerful. Unlike the conventional models, strained or crowded systems such as cyclopropane are no more difficult to construct than normal molecules. Geometrical parameters (bond lengths and angles) may be measured allowing quantitative comparisons of molecular structure. In addition to skeletal models which, like plastic models, are intended to depict bonding, SPARTANBuild is able to portray molecules as space-filling models, thereby conveying a sense of overall size and shape. Finally, SPARTANBuild allows refinement of molecular structure with "molecular mechanics," and reporting of strain energies. Thus, "qualitative impressions" which might follow from inspection of plastic models can be quantified using SPARTANBuild.

The greatest attribute of SPARTANBuild models is actually shared by plastic models. It is simply the ability to portray molecules as three dimensional, and ask questions about molecular structure and chemical reactivity in this context. This offers great advantage over two dimensional drawings.

This tutorial will show you how to : i) build molecules from atoms, functional groups and rings, and ii) minimize their energies.

SPARTANBuild is very similar in appearance to SPARTANView, and many of its operations (moving and rotating molecules, measuring distances and angles) and model displays are the same. It differs from SPARTANView in that it incorporates a set of "building blocks" from which molecules can be constructed, a set of specialized building tools and an energy minimizer to produce a refined structure.

Building a Molecule from Atoms

A simple way to build a molecule is from its atoms. For example, pent-2-yne-4-ene can be assembled from five "atoms," sp^3C, spC, spC, sp^2C and sp^2C.

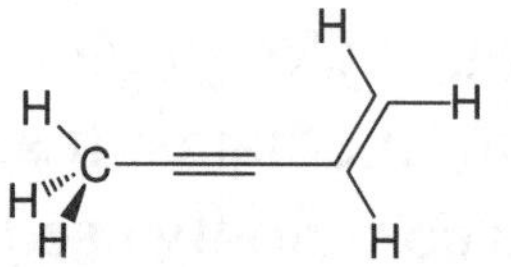

Start SPARTANBuild	Starting the program opens a large SPARTANBuild window (this is blank initially), a model kit, and a toolbar. Models are assembled in the window.
Start building pent-2-yne-4-ene	
1. *Click* on  in the model kit.	The  button becomes highlighted.
2. *Click* anywhere in the window.	An sp^3 carbon atom with four unfilled valences (white) appears as a Ball and Wire model.

Note that sp^3 carbon is only one of the five kinds of carbon atoms available in the model kit. Each of these carbons (sp^3, sp^2, sp, aromatic and trigonal) is characterized by a specific number and kind of unfilled valences (used to make bonds) and an idealized geometry. For example, sp^3 carbon has four unfilled valences, any or all of which can be used to make single bonds to other atoms. Any unfilled valences which remain when you finish building are turned into bonds to hydrogen.

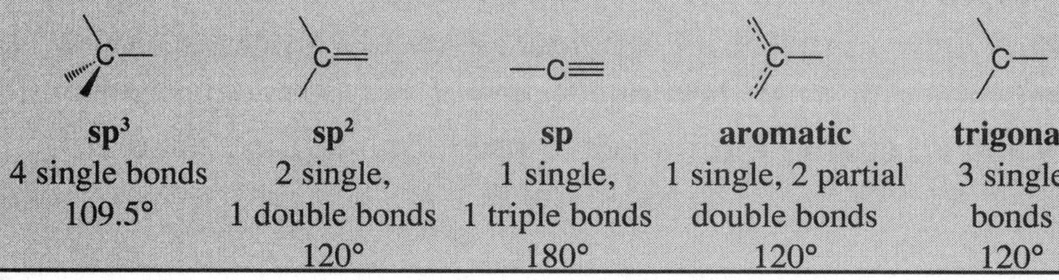

sp^3	sp^2	sp	aromatic	trigonal
4 single bonds 109.5°	2 single, 1 double bonds 120°	1 single, 1 triple bonds 180°	1 single, 2 partial double bonds 120°	3 single bonds 120°

Finish building pent-2-yne-4-ene	
3. *Click* on spC in the model kit.	This selects the carbon atom with one single and one triple unfilled valences.
4. *Click* on the tip of the unfilled valence on the carbon in the window.	This makes a carbon-carbon single bond.
5. *Click* on the tip of the triple unfilled valence of the sp carbon in the window.	This makes a carbon-carbon triple bond.
6. *Click* on sp^2C in the model kit.	This selects the carbon atom with two single and one double unfilled valences.
7. *Click* on the tip of the unfilled valence on the sp carbon in the window.	This makes another carbon-carbon single bond. Bonds can only be made between unfilled valences of the same type.
8. *Click* on the tip of carbon's double unfilled valence in the window.	This makes a carbon-carbon double bond and completes the model.

Building a Molecule from Groups and Rings

SPARTANBuild simplifies construction of molecules containing organic functional groups and rings by providing a library of "pre-built" structures. For example, phenylacetate can be built using the "carboxylic acid" functional group and the "benzene" ring.

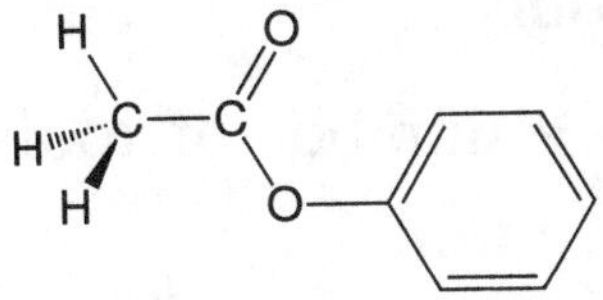

Select Clear from the Edit menu	This removes the existing model from the window.
Build Phenylacetate	
1. *Click* on sp^{3}C in the model kit, then *click* in the window.	This places an sp^3 carbon on screen.
2. *Click* on the **Groups** button in the model kit.	This indicates that a pre-built functional group is to be used.
3. Select **Carboxylic Acid** from the **Groups** menu.	An icon of the carboxylic acid group appears in the model kit.
4. Examine the unfilled valences of the carboxylic acid group and find the one marked by a small circle. If necessary, *click* on the group's icon to move the unfilled valence to carbon.	This group has two different unfilled valences that can be used to connect it to the model. The "active" unfilled valence, indicated by a small circle, can be changed by *clicking* anywhere on the group's icon in the model kit.
5. *Click* on one of the unfilled valences of the sp^3 carbon in the window.	This bonds the carboxylic acid group to the sp^3 carbon.
6. *Click* on the **Rings** button in the model kit.	This indicates that a pre-built ring is to be used.
7. Select **Benzene** from the **Rings** menu.	An icon of the benzene ring appears in the model kit.
8. *Click* on the unfilled valence on oxygen of the carboxylic acid group.	This adds the benzene ring to your structure, leaving you with phenylacetate.

Specialized Building Tools

SPARTANBuild provides a number of tools to assist in molecule building.

Make Bond

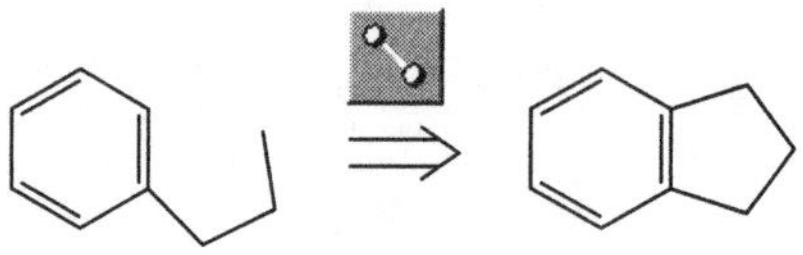

Break Bond

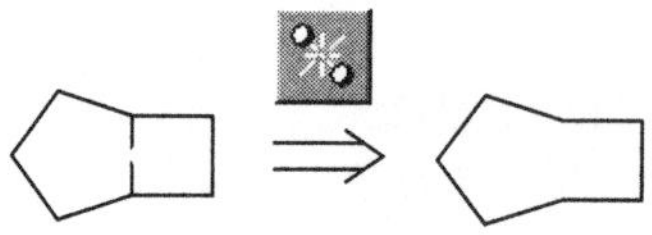

Delete 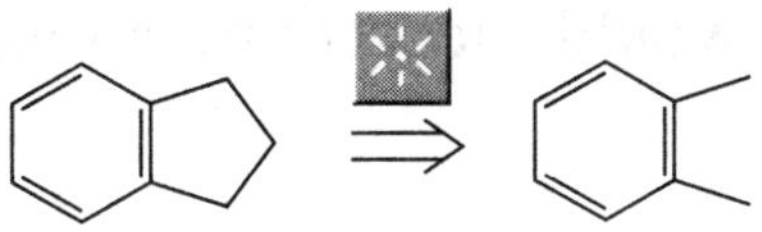

Internal Rotation 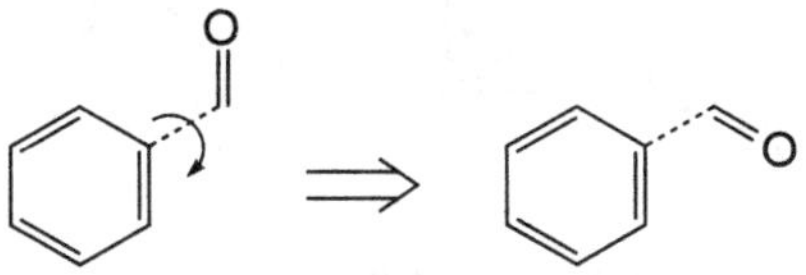

Atom Replacement 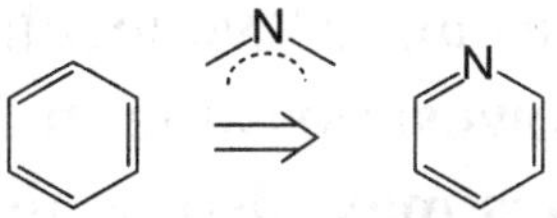

MAC

	*Click* on two unfilled valences to make bond.
	*Click* on a bond to break.
	*Click* on atom or unfilled valence to delete.
Internal Rotation	*Click* on bond to select. Hold down both **space bar** and the button and move mouse.
Atom Replace-ment	Select atom from model kit and *double-click* on an atom in the model.

PC

	*Click* on two unfilled valences to make bond.
	*Click* on a bond to break.
	*Click* on atom or unfilled valence to delete.
Internal Rotation	*Click* on bond to select. Hold down both **Alt key** and the left button and move mouse.
Atom Replace-ment	Select atom from model kit, and *double-click* on an atom in the model.

Lowest-Energy Structures

Building will not always lead to the best possible (lowest-energy) molecular structure. Where bonds are made or broken or where atoms are deleted, the resulting geometry may be highly distorted. For example, cycloheptene constructed by joining the terminal carbons in *cis*-2-heptene would incorporate an unrealistically long carbon-carbon bond.

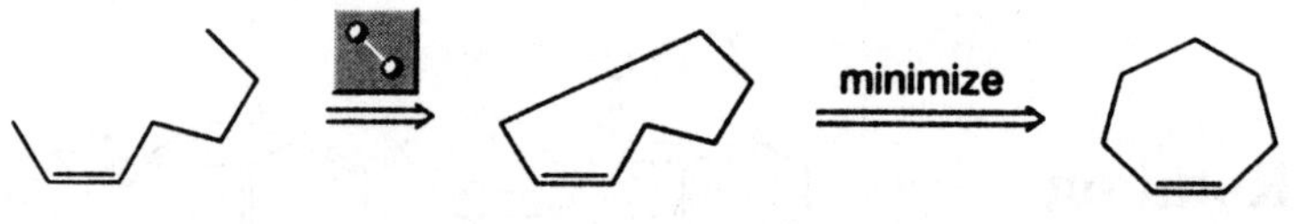

SPARTANBuild provides a **Minimize** tool for such a situation. This invokes a procedure called "Molecular Mechanics" to search for the structure with the lowest possible strain energy. Strain energies from molecular mechanics may also be used to compare the energies of either stereoisomers or of conformational isomers, for example.

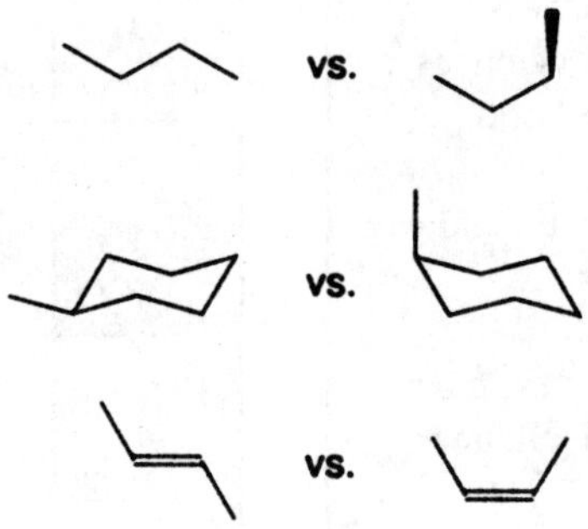

Index of Models

E

F

G

H

I

L

M